Wild Flowers
of
NEW ZEALAND

OWEN BISHOP
Photographs by Nic Bishop

Hodder & Stoughton
AUCKLAND LONDON SYDNEY TORONTO

First published 1990
ISBN 0 340 533234

Typeset by Glenfield Graphics Ltd, Auckland.
Printed and bound in Hong Kong for Hodder & Stoughton Ltd,
46 View Road, Glenfield, Auckland, New Zealand.

Contents

List of colour plates

Introduction

This book describes over 400 of the flowering plants that you are most likely to notice while you are out and about in New Zealand. The book is arranged so that it can be used by a beginner, yet contains enough information to satisfy the enthusiast.

The wild flowers of New Zealand comprise native plants and introduced plants. The native plants are mostly unique to New Zealand, and have existed here for millions of years. The introduced plants have been brought to New Zealand from almost all parts of the world, mainly during the 19th and 20th centuries. Many of the introduced plants were brought here deliberately, generally from Europe, for cultivation in farms and gardens. Since then they have escaped and become naturalised on roadsides, on waste land, in scrub or in similar situations. Other plants have been introduced accidentally, as weed seeds in soil attached to imported plants, for example, or as impurities in crop seeds. Whether introduced deliberately or accidentally, these plants are now a permanent part of the New Zealand scene; they tend to be the ones we notice most often, for they inhabit areas in and around our settlements. Wherever the native vegetation has been cleared for cultivation, for making roads or for building, these 'opportunists' have had their chance to become established. They have invariably taken that chance and have usually succeeded. This book includes the commonest of the introduced plants that you are likely to find in such areas – the ones you are most likely to see in everyday living.

The book includes common native plants too, but to find these you will usually need to travel beyond the areas of cultivation and settlement. In reserves (both local and national) and other relatively undisturbed places there is a profusion of native plants waiting to be found and identified. If you go further, to the wilder areas of New Zealand's national parks, you will find very many more. But we have had to be realistic and recognise that the average reader is not likely to get so far. Accordingly, apart from a few species of special interest, we have not included plants that are found only in the remoter areas or on the offshore islands. We suggest that wilderness trampers and mountaineers look in the book list (p. 182) for suggestions on specialist books to take with them.

The plants illustrated in this book are mainly herbs, that is to say, they are non-woody plants. We have also included a few small flowering shrubs that are very common and are only slightly woody. As there are already several excellent books for identifying trees (see the book list), we decided not to include any trees here. When selecting which plants to include, we favoured those which are widespread in both the North Island and the South Island. As well as these, we have added some that are particularly common in one island, though rare or absent from the other. The grasses, sedges and rushes are very difficult groups for the non-specialist to identify, but we have included a key to tell you which group is which. (A detailed book on introduced grasses is listed on p. 182.)

The photographs present flower portraits that show the characteristic features of each plant to the best advantage. Their aim is to help you find the names of the plants with the minimum of trouble. If you want to, just flick through the pages until you come to a picture that looks like the flower you have found.

If you are more systematically inclined, identification keys and the order of arrangement of the photographs will lead you quickly to the picture of the flower you are trying to name. Accompanying each picture is a brief description of the distinctive features of each kind of plant; this helps to tell you whether or not your identification is correct. In some instances the descriptions also refer to related plants of similar appearance, showing you how to distinguish them from the plant illustrated.

CLASSIFYING PLANTS

There are several thousand different kinds of flowering plants living in New Zealand. If we are to give them names, it makes sense to group similar plants together – to classify them. This section describes how plants are classified.

When we speak of a 'kind of plant' we usually mean what biologists call a species. A species (the plural is also species) is something that we can recognise as being distinct in one or more ways from all other species. As an example, we can take great bindweed (p. 103), with its creeping, twining stems and large, white, funnel-shaped flowers. It is a common and distinctive inhabitant of hedges and forest margins on both islands. It is obviously not the same species as pink bindweed (p. 103), which has smaller pink or pink-and-white flowers and other distinctive features. Great bindweed differs even more from shore bindweed (p. 103) which has the smallest pink-and-white flowers and distinctively shaped fleshy leaves. Once you have been shown the differences between these three kinds of bindweed, you have no difficulty in deciding which is which. Each of these three bindweeds is a distinct and easily recognisable species.

The difference between one species and another goes even deeper than appearances:

biologists define a species as a group in which members of the group are capable of breeding with each other, but not with members of other species. This is the criterion that biologists use when deciding whether two kinds of plant belong to the same species or not. Unfortunately, this strict definition is not always easy to apply, as it may sometimes happen that two species are found to be capable of cross-breeding with each other – of producing hybrids. This is just an instance of the way in which our neat ideas about classification can be upset by geneticists. The exact status of many of our so-called species is always coming under review as more is discovered about their genetic make-up and breeding behaviour. Yet, despite the exceptions and the complications, the idea of a species as being a self-breeding unit remains a useful one.

The three species of bindweed have several features in common, including twining stems and funnel-shaped flowers; they belong together. People have recognised this connection by giving them all the same name – bindweeds. They differ as a group from other natural groups such as buttercups, clovers, and speedwells.

A natural group of plants such as this is called a genus (plural genera). The word genus simply means 'a kind': the three bindweeds belong to the genus *Calystegia*, clovers belong to the genus *Trifolium* and speedwells belong to the genus *Veronica*. But one name is not enough, and the scientific name of a plant usually consists of two names. First we have the name of its genus, so the names of all three bindweeds begin with *Calystegia*. We then indicate which of the three bindweeds we are talking about by adding a second name; this is the name of the species. Great bindweed is *Calystegia silvatica*, pink bindweed is *Calystegia sepium* and shore bindweed is *Calystegia soldanella*. Thus the scientific name of a plant consists of its genus name followed by its species name.

Some genera have many species: for example there are 53 species of the buttercup genus *Ranunculus* living in New Zealand, and over 400 species world-wide. Some genera have only one species, for example, shamrock pea, *Parochetus communis* (p. 114), is the only member of its genus. It is a plant that stands out on its own, and is easy to identify.

A group of genera that have several features in common is known as a family. For example, the bindweeds' genus (*Calystegia*) may be grouped with a slightly different kind of bindweed genus (*Convolvulus*, p. 102), the morning glory genus (*Ipomoea*, p. 102) and some less common genera. All of these genera are twining plants with flowers of similar structure. Collectively they are known as the Bindweed Family.

There are several hundred families of flowering plants, including the Bindweed Family, the Buttercup Family and the Rose Family. Each family has a scientific name; this is made up by taking the name of a typical genus of the family, possibly shortening it or altering one or two of its ending letters, and then adding '-aceae'. For example, the Bindweed Family is named after the typical genus *Convolvulus*, so it is called the Convolvulaceae. Similarly, the Iris Family is named after the typical genus *Iris* and is called the Iridaceae. Wherever you see the ending '-aceae', it refers to a plant family.

Finally, the families may form larger groups, known as classes. As explained on p. 17, we group the families of flowering plants into two classes, called (for short) monocots and dicots.

FINDING YOUR WAY

To identify a plant quickly you need to know your way about this book. We have adopted a system that should be easy to understand. First of all, the photographs and descriptions are set out under the two main classes:

- Monocots – begin on p. 19
- Dicots – begin on p. 43

Under each class, the descriptions are grouped by families, in alphabetical order. For each family there is a description of its main features, as explained below.

Under each family, the genera are listed in alphabetical order of genus name. Under each genus, the species are listed in alphabetical order of species name. In this way, related plants of similar appearance are close together in the book.

Another way of finding a description or photograph quickly is to use the index, where all species are listed under their species and their common names.

FAMILY FEATURES

Each family description begins with a set of letters to indicate the main family features. See the inside of the front cover for an explanation.

Following the coded information is a summary of the family, including the names of members (not necessarily found in New Zealand) that are of economic importance or of special interest for any other reason.

DESCRIBING THE PLANTS

The descriptions begin with the scientific name of the plant, followed by the common name or names, if any. When a species is first identified in the text an asterisk (*) after the scientific name indicates that the species is considered to be an introduced member of the New Zealand flora.

The descriptions of the plants are as short as possible, so that you can read through them quickly. They are intended to be used for confirming identity, not as complete botanical descriptions, so they emphasise those features that help you to distinguish between the species and any other common species that has a similar appearance. If there are any technical words in the descriptions that you do not understand, refer to the inside of the back cover.

When identifying certain plants it is necessary to check features of the plant as a whole, as well as details of its flowers. A single photograph cannot easily cover large-scale as well as small-scale features. For most species, the photographs show the flower details, and the descriptions deal with any large-scale features that are important for identification.

Common names and scientific names

Although, as in the example of the bindweeds, the common names of plants may correspond to a natural group, you cannot rely on this. People have often given the same name to plants that are *not* closely related. For example, the name 'bachelor's buttons' is given to two quite unrelated New Zealand species, as well as to a third species common in England.

This is why it is essential that plants have scientific names too. Scientific names take a little getting used to, but they have advantages:

- Plants of a similar kind have the same genus name. With a little experience, you soon get to know the main features of the common genera, and you soon get to know the common genera in a family. In your mind, you build up a picture of the common plants, sorted into groups that have similar features; this helps a lot in learning to recognise plants, or in using books like this one. The common names are not nearly as helpful and, as in the example of bachelor's buttons and many others, may be positively misleading.

- Common names for the same plant may differ in different countries or even in different parts of the same country. Conversely, the same common name may be used for two or more different plants. For example, *Verbascum blattaria* is called white mullein in New Zealand, but it is called moth mullein in Europe. This is not all: the same name, moth mullein, is used in New Zealand too – for an entirely different species, *Verbascum virgatum*. Back in Europe, this species is called twiggy mullein! There seems to be no end to the confusion.

- Scientific names are international. The common names of species are bewildering to visitors from other countries. Since this book is written for tourists as well as for residents, we have named all species in this book by their international scientific names as well as by their common New Zealand names.

SIZE OF FLOWERS

One feature that is not easy to show precisely in a photograph is size. Sizes of flowers are indicated by the figures on the right-hand side of the page, level with the common name. This usually gives the diameter in millimetres. If flowers vary in diameter, the lower and upper limits are given, for example, '15–25' means that the flowers vary in diameter from 15 mm to 25 mm. For a few species, the length of the flower is easier to measure, and we quote the length followed by 'long', for example, '50 long'.

HEIGHTS OF PLANTS

The height of mature plants is indicated by using standard terms:

Height	*Description*
Up to 100 mm	**low**
100 to 300 mm	**short**
300 to 600 mm	**medium**
Over 600 mm	**tall**

Some species span more than one height group; for these plants we use terms such as 'medium to tall', meaning that plants range in height from 300 mm to over 600 mm.

Each description is completed by stating the distribution, habitat and flowering dates of the species. Checking that the plant you have found is supposed to grow where you found it, and is supposed to be flowering on the date you found it, is a valuable way of confirming its identity.

Most of the plants described in the book are common in both the North and South Islands. To save space, we mention distribution only when a species is restricted to one island or a part of an island. The habitat of the plant is described in general terms. It is sometimes possible to find a plant living in some other habitat, though perhaps not for long. Flowering dates are given where these are known; plants may be found in flower slightly earlier than these dates in northern areas, and slightly later in southern areas. In sheltered areas, certain plants may be found in flower well outside the range of dates specified.

The description is often followed by interesting points about the uses or history of the plant.

USING A KEY

Although it often takes only a few seconds to leaf through the book and pick out a particular photograph, there are occasions when this approach does not work well. For example, there are many species that have yellow, dandelion-like flowers. The plants look alike at first glance, and so do the photographs. To help you find your way to the correct photograph, we have included keys to these groups: a key points out to you the important features for distinguishing between different members of a group.

The keys used are two-choice keys. At each stage you have to choose between two alternative descriptions. As an example, here is the first part of the key to yellow, dandelion-like flowers (the rest is on p. 79):

1 Flower-head solitary on an unbranched stem ▸ 2
Flower-heads more than one on a branched stem ▸ 5

2 Leaves simple, with shaggy white hairs ▸ ***Hieracium pilosella*** (p. 81)
Leaves lobed or toothed ▸ 3

To use this key, begin at line one and consider the two alternative descriptions. Does your plant have solitary flower-heads on unbranched stems? If so, follow along to the end of the line and note the number '2'. If it has several flower-heads on a branching stem, note the number '5'.

Assuming that the plant you are trying to identify has solitary flower-heads on an unbranched stem, you consider the pair of alternatives numbered '2': you are now asked to decide if the plant has simple leaves, with shaggy white hairs. If so, there is only one plant in the key answering to this description. Its scientific name, *Hieracium pilosella*, is given at the end of the line. If the plant does not have simple leaves, but has lobed or toothed leaves instead, there are a few more pairs of alternatives to consider, so you go on to '3'. Following through from one choice to the next quickly leads you to the name of the plant.

Keys are easy to use, but remember that a key often includes only the commonest plants. It may let you down if the plant you are trying to identify is not included in the key. Sometimes you may realise this when you find that your plant fits neither of the alternative descriptions.

When you have run through the key, check your specimen against a photograph or a description, as these give positive proof that the identification is correct. On reaching '*Hieracium pilosella*', for example, look at the picture on plate 12 and read the description: does your specimen look like the one in the photograph? Does it have runners? Does it have a greyish white felt beneath the leaves? Does it have brittle hairs on its fruits? If the answer to all these questions is 'Yes', the identity is confirmed. If not, you possibly made a mistake in using the key, or perhaps the key does not include the plant you are trying to identify.

We have included keys to various groups of plants, as listed in the Contents. One of these is a key to the families of monocots. It is not practicable to include a key to the dicot families, because such a key would depend a lot on features that require detailed examination, often with a microscope. Instead we give a table of family features that are readily determined in the field (p. 45); a computerised version of this appears on p. 176.

The flowering plants

The flowering plants are divided into two classes, monocotyledons and dicotyledons; these are usually known as 'monocots' and 'dicots', for short. The main differences between monocots and dicots are:

Monocots:

leaves have parallel veins

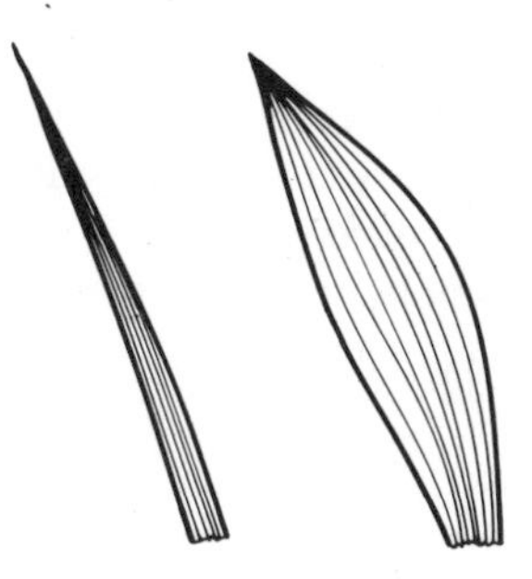

flower-parts are usually in threes

Examples: lilies, orchids, grasses, irises.

Dicots:

leaves have a main vein and a branching network of veins

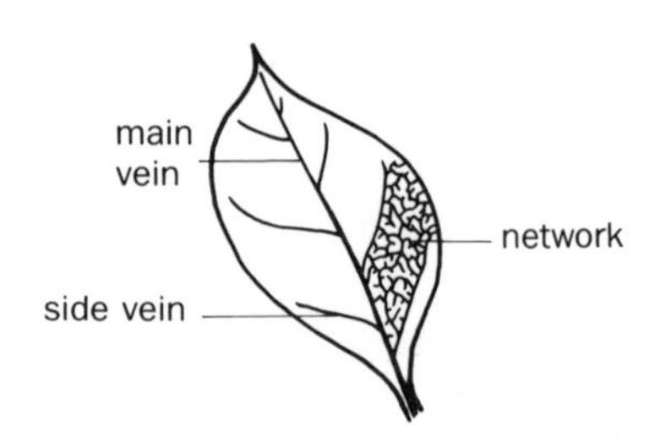

flower-parts are usually in fours or fives

Examples: buttercups, violets, daisies.

Another difference between these two classes is that the embryos in the seeds of the monocots have one seed-leaf (cotyledon), whereas the embryos of the dicots have two seed-leaves. This difference is so important to botanists that it gives the classes their names, but it is not a difference that it is easy to use for identifying plants in the field.

MONOCOTS

This class consists mainly of herbaceous (that is, non-woody) plants, but it also contains a few trees, such as the palms and cabbage trees. It includes plants that reproduce by bulbs and corms, many of which have bright showy flowers: daffodil, tulip, crocus, watsonia, gladiolus, snowflake, bluebell, hyacinth, day lily, crocosmia, agapanthus, the orchids and many other decorative garden plants, some of which are also found growing wild in New Zealand.

The grasses stand out as some of the most successful and widespread of all monocots. Apart from the wild grasses, there are the crop grasses, which provide food for humans directly, or indirectly as fodder for livestock. Important food grasses include:

- Wheat – the most important world-wide, and the most important in temperate climates.
- Rice – the second most important world-wide, especially in Asia.
- Sugar cane – provides more than half the world's sugar.
- Other important grasses, such as barley, oats, maize, and millet.

As well as the grasses, the monocots include other important food plants, such as the palms (oil palm, coconut palm, date palm, sago palm and sugar palm), taro, yam, and Chinese water-chestnut.

IDENTIFYING MONOCOTS

Follow this key (see p. 14) to find out the family to which your specimen belongs:

1 Grass-like plants, with narrow, sheathing leaves ▸ 2
Not as above ▸ 4

2 Leaves have a distinct 'joint' with a 'tongue' at the joint
▸ Grass Family **Poaceae***
Leaves without distinct joint ▸ 3

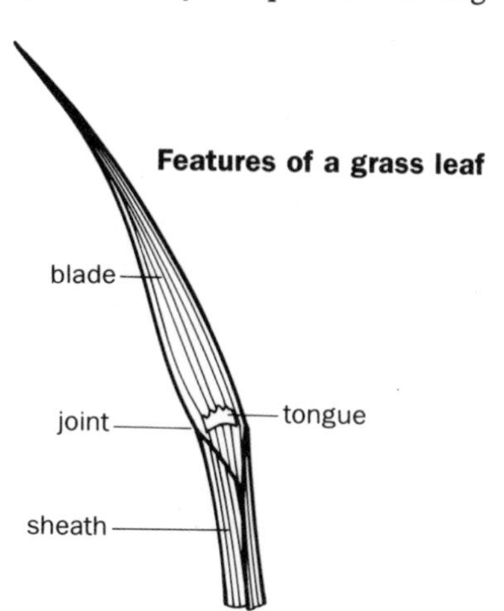

Features of a grass leaf

* The Grass Family is called the Graminae in some other books.

3 Flowers with six greenish or brownish tepals ▶ Rush Family **Juncaceae**

Flowers with brownish or greenish scales or hairs instead of tepals, flowering stem hollow and usually triangular in section ▶ Sedge Family **Cyperaceae**

4 Flowers very small, often without sepals or petals, clustered on a thick, fleshy stalk, the inflorescence enclosed in a large, leaf-like white or coloured sheath ▶ Arum Family **Araceae** (p. 23)

Not as above ▶ 5

5 Flowers with three coloured or white petals and three green sepals ▶ Spiderwort Family **Commelinaceae** (p. 25)

Not as above ▶ 6

6 Flowers with three stamens ▶ Iris Family **Iridaceae** (p. 27)

Not as above ▶ 7

7 Ovary superior ▶ 8

Ovary inferior ▶ 13

8 Plant reproducing by bulbs (i.e. like an onion) ▶ 9

Plant without bulbs ▶ 11

9 Leafless stalk topped with flowers (possibly basal leaves may sheathe the stalk but they do not grow from it) ▶ 10

Leafy flower-stalk; petals often patterned or spotted and curling back ▶ Lily Family **Liliaceae** (p. 31)

10 Inflorescence is an umbel, with (usually) two membranous bracts immediately below it; plant may have onion smell ▶ Onion Family **Alliaceae**** (p. 22)

Inflorescence is a raceme or spike, each flower with its own bract; no onion smell ▶ Hyacinth Family **Hyacinthaceae**** (p. 26)

11 Leaves and flowering stalks grow from thick, fleshy roots ▶ Asphodel Family **Asphodelaceae**** (p. 24)

Leaves and flowering stalks grow from rhizomes ▶ 12

12 Leaves in two opposite rows with sheathing bases (equitant); branching inflorescence ▶ Harakeke Family **Phormiaceae** (p. 40)

Strap-shaped leaves in a basal tuft; blue or white flowers in umbels at the top of a long, leafless stem ▶ Onion Family **Alliaceae**** (p. 22)

** Plants in these families are all included in the Lily Family in some books, but are now considered to belong to separate families.

13 Sepals joined to make a narrow three-toothed tube, petals similarly joined ▶ Ginger Family **Zingiberaceae** (p. 42)

Sepals and petals usually not joined, and the tube, if any, not narrow ▶ Orchid Family **Orchidaceae** (p. 32)

The flowers of the grasses, and also the rushes and sedges, are relatively small. It is difficult for anyone but an expert to identify members of these families, which is why we do not go into further details of them in this book. For the other families listed in the key, turn to the page indicated and study the photographs.

Onion Family – Alliaceae

H B E – R
P 3+3 or (3+3)
M 3+3
F (3)

These are perennial plants, which have a bulb or bulb-like corm; a few have rhizomes. The leaves are narrow and basal. They sheathe the flower-stalk and so may appear to be alternate. The inflorescence is an umbel, and there are two or more large membranous bracts at the base of the umbel (sometimes only one bract).

Most species of onion (*Allium*) and many other members of the family have a characteristic 'onion' smell when crushed. The onions and similar plants, including leek, shallot, chive and garlic, are minor but commercially important food crops, more often used to provide flavouring than for their nutritional content. Some other members of the family, such as agapanthus and a few *Allium* species, are grown in gardens for ornament.

Allium triquetrum★ ★ = introduced species — Plate 1

THREE-CORNERED GARLIC — 15

A medium perennial, growing from bulbs. The plant has a distinct and rather unpleasant garlic-like smell. The leaves are long and narrow and solid; they are easily crushed, producing the strong-smelling juice. The leaves arise only from the base of the plant, sheathing the inflorescence stalk. The name 'three-cornered' comes from the stalk of the inflorescence, which is triangular in section, with a pronounced 'wing' at each corner. The inflorescence bears a dense cluster of about 3–15 white flowers on short stalks. As the flowers open and these stalks lengthen, the flowers droop, as seen in the photograph. The inner surface of each tepal has a central green stripe.

This is a common weed, found in gardens and waste places. It flowers in October and November.

It may be confused with the wild onion, *Allium vineale*★, but this has hollow leaves that are more or less circular in section. The inflorescence stalk is circular in section, too. Wild onion usually produces small bulbs (bulbils) instead of flowers. In Europe the wild onion is a serious weed of arable land; it is eaten by livestock and, though it does not poison them, it causes unpleasant flavours in meat and milk.

Arum Family – Araceae

H B EC – R
P usually 0
M up to 6
F (1 to 3)

The inflorescence is the distinctive feature of this family; in most species it consists of a vertical, fleshy stalk (spadix) densely covered with small flowers. The inflorescence is usually surrounded, wholly or partly, by a leaf-like spathe which may be white, green or brightly coloured. The end of the spadix is often enlarged, without flowers and has glands that produce strong-smelling (often unpleasant-smelling!) substances, which attract the insects that pollinate the flowers.

This family includes an important food plant, the taro (*Colocasia esculenta*), which is occasionally found growing wild in New Zealand. The rhizomes of this plant are starchy and edible. Its European relative, *Arum maculatum*, variously known as lords and ladies, cuckoo-pint and parson-in-the-pulpit, has rhizomes that are ground to produce Portland arrowroot, an easily digestible form of starch. Many other genera are grown as garden plants and house plants, and one of the commonest of these is the swiss cheese plant (*Monstera deliciosa*), with its characteristic 'holey' leaves.

*Zantedeschia aethiopica** — Plate 1

ARUM LILY — **250 long**

A tall, robust, perennial plant with arrow-shaped, dark-green leaves up to 250 mm long. The veins of the leaf are fine and a slightly lighter green than the rest of the leaf. The spathe is white, up to 250 mm long, and the upper region of the spadix is yellow.

It lives in damp places, often in damp areas of pasture, and flowers from October to December.

The species name *aethiopica* suggests that this plant originated in Ethiopia: in fact, it comes from South Africa. The name *aethiopica* was given to it in the 18th century, when Africa was little known to Europeans, and at that time the name Ethiopia was used for most regions of Africa.

Asphodel Family — Asphodelaceae

HfewST A Si – IR
P 3+3 inner tepals may differ from outer tepals
M 3+3
F (3)

The leaves are arranged spirally at the base of the plant, often thick and juicy. A very familiar member of this family is the garden plant, red-hot poker (*Kniphofia*).

Asphodelus fistulosus★ **Plate 1**

ASPHODEL **20**

A medium, perennial plant with fleshy, tufted leaves arising only at the base of the plant; the leaves are 120–250 mm long, hollow, and almost circular in section. The inflorescence is on a stalk 200–600 mm long and bears numerous white flowers. Each tepal has a central stripe, either green or purple. The filaments of the stamens are swollen and have hairs at their bases.

This plant is usually found near the sea, in coastal scrub or waste land or growing between low-tide and high-tide levels; it flowers from July to September.

Bulbinella gibbsii **Plate 1**

MAORI ONION **10–14**

A short to medium, perennial plant. The leaves are up to 3 cm wide and all grow from the base of the stem. They die back in autumn and are renewed in spring. The inflorescence is a raceme of yellow flowers, the stalk of the inflorescence being visible between the flowers. The filaments of the stamens are without hairs. *Bulbinella gibbsii* is distinguished from other species of maori onion found in New Zealand by the fact that the tepals remain yellow, smooth and erect around the maturing fruit. This can be seen in the bottom flower at the front of the left-hand plant in the photograph. The only other species with this feature is restricted to Auckland Island and Chatham Island; in other species, the tepals wither and hang down. Another distinguishing feature is that the older leaves arch over, instead of remaining more or less erect.

There are two varieties of *Bulbinella gibbsii*. The photograph illustrates *B. gibbsii* var. *balanifera*: the fruit is brown and, with the shrivelled remains of the stigma on top of it, looks like a small, ripe acorn. The bracts at the base of the inflorescence can be clearly seen to be much shorter than the flower stalks. In the other variety,

B. gibbsii var. *gibbsii*, found only on Stewart Island, the fruit is more globe-shaped and the bracts are at least as long as the flower-stalks.

Found in well-drained montane and subalpine areas of the North and South Islands: flowers in January and February.

Bulbinella hookeri — Plate 1

MAORI ONION — **14**

A short to medium, perennial plant. The leaves are up to 30 mm wide, all arising from the base of the stem. The leaves remain erect, not arching over as in *B. gibbsii*, and they die back in autumn and are renewed in spring. The inflorescence is a raceme of yellow flowers, the stalk of the inflorescence being visible between the flowers. The filaments of the stamens are without hairs. The tepals wither and hang downward at fruiting; the lowest flowers in the young inflorescence of the photograph are beginning to show this. Another distinguishing feature is that the ovary, and later the fruit, is on a short, narrow stalk within the flower.

Found in damp, shady areas in montane to low-alpine tussock grassland, it flowers from November to January.

Spiderwort Family – Commelinaceae

H A E - R
K 3
C 3 white, purple or blue
M 3+3 may be reduced or absent; hairy filament
F $\underline{(3)}$

These are usually short to medium-sized, fairly succulent herbs, the stems of which have a characteristic 'jointed' appearance. The colourful flowers and the ease with which they reproduce from stem cuttings have earned many members of this family a place in gardens, and in homes as ornamental plants.

*Tradescantia fluminensis** — Plate 3

WANDERING JEW — **20**

A medium, perennial herb, its trailing stems cover the ground in dense patches, taking root at intervals. The leaves are broad and oval, but the parallel veins confirm that this is a monocot, as can be seen in the photograph. The bases of the leaves curl round to form sheaths surrounding the stem.

It is found in damp, shady places, often overgrowing the local vegetation: flowers in December and January.

The name 'wandering Jew', descriptive of the ease with which this plant spreads, is also applied to some other members of the family. The common house plant, *Zebrina pendula*, with purple flowers and leaves often striped white, green and purple, is another wandering Jew. The hairs on the stamens make an interesting subject for a low-power microscope: if mounted in water, the hairs can be seen to consist of exceptionally large cells. It is usually possible to see the living material inside the cells (nucleus and cytoplasm) slowly and continuously circulating.

Hyacinth Family – Hyacinthaceae

H B E – R
P 3+3 sometimes free
M 3+3
F <u>(3)</u>

These are perennial herbs with bulbs, and narrow, entire, sheathing leaves at the base of the inflorescence stalk. The inflorescence is a raceme or spike, usually each flower being in the axil of a narrow bract. The flowers may have a wide range of colours.

The best known members of this family are those grown in gardens for their showy inflorescences. These include hyacinth, grape hyacinth, Star of Bethlehem and bluebell, all of which (except the hyacinth) have escaped and now grow wild in New Zealand.

Scilla non-scripta★ **Plate 1**

BLUEBELL **15–20 long**

A short to medium perennial with long, narrow leaves growing from a bulb. The inflorescence is a short raceme of a few blue or white bell-shaped flowers. The tube of the flower is short and the ends of the six more or less identical tepals are curled back. The filaments of the six stamens, the stalks of the flowers, and the bracts at the base of the flower stalks are all the same colour as the flower.

A garden escape that has become well established in grassy roadside areas: flowers in September and October.

The bluebell was introduced from Europe, where, in late spring, it carpets large areas of deciduous woodland with its nodding, blue flowers. A related species, introduced into New Zealand from the Mediterranean region, is *Scilla peruviana*★, which has inflorescences of smaller but more numerous flowers. *S. non-scripta* has both a bract and a bracteole (smaller) at the base of each flower stalk, as seen in the photograph, but *S. peruviana* has only a bract. The Elizabethans obtained starch from the bulbs and used it to stiffen the ruffs they wore around their necks. The sticky fluid from the stalks was used for gluing the pages of books and for fixing flight feathers to arrows.

Iris Family – Iridaceae

H Eq E – RI
P (3+3) or 3+3 petal-like
M 3
F (3) three-lobed style, long and narrow in some species, petal-like in some other species

These are almost all perennial herbs. Although they produce flowers and seeds, many rely on their corms or rhizomes for reproduction. The three petals may look almost the same as the three sepals, so that the flower appears to have six petals, but in some species the sepals and petals differ in colour, size or shape. The leaves are usually all at the base of the stem, long and narrow, sheathing and overlapping (equitant, see the inside of the back cover).

Many members of this family originate from South Africa, so the species found wild in New Zealand are mostly garden escapes. However, there is one species, *Libertia*, which is thought to be native. One member of this family, the saffron (*Crocus sativus*), was widely cultivated in medieval times as a source of an orange dye; the colouring was prepared from the orange stigmas of the flower, as many as 4300 flowers being needed to produce one ounce of saffron. The product was expensive and saffron cultivation was an important industry, so much so that the plant has given its name to the town of Saffron Walden in England.

Aristea ecklonii★ — Plate 2

ARISTEA — **20**

A medium, perennial plant with a fan of narrow leaves, which are reddish in colour at the base. The flowers are in clusters of five to seven, but an individual flower lasts for only a few hours, so there is usually not more than one flower open, in each cluster, at any one time. The petals are broader than the sepals, but similar in length and colour.

It is found on roadsides in Northland and Auckland city: flowers in October.

Crocosmia ×crocosmiiflora★ — Plate 2

MONTBRETIA — **40–50**

A medium to tall, hairless plant forming dense clumps. The leaves are flat. The inflorescence is 150–300 mm long, consisting of a one-sided spike, slightly branched or unbranched, with a zig-zag stalk. Flowers are tubular at the base, with six lobes equal in length. The three branches of the style are three-lobed at their tips.

It is found on roadsides, grassy areas and waste land, abundant in places: flowers from December to February.

The '×' in the name of this plant indicates that it is a hybrid species. Having been cross-bred and becoming popular in gardens, it has proved itself to be particularly successful in the wild. Where garden refuse has been dumped by the roadside, montbretia has established itself and spread rapidly. A related species is *Crocosmia paniculata*★, occurring in the South Island (Nelson and Westland); this is identified by its folded or pleated leaves, its branched inflorescence and its smaller (15 mm diameter) orange flowers, tinged with crimson.

Gladiolus undulatus★ — Plate 2

WILD GLADIOLUS — **80**

A medium, perennial plant, producing small corms. The leaves are green with a purple tinge and the leaf sheaths are reddish purple. This species is easy to identify by its flowers, with their greenish-yellow, pointed-tipped, wavy-edged petals. The three lower lobes of the flower may have central purple stripes.

It is common on roadside verges, vacant sections and waste places in Northland and Auckland city: flowers in December.

The name *Gladiolus* is the Latin for a small sword, referring to the shape of the leaves.

Libertia ixioides — Plate 2

MIKOIKOI – TUKAUKI – MANGA-A-HURI PAPA — **15–20**

A medium to tall perennial with green leaves, tending to become yellow where they are exposed to sunlight. The veins of the leaves may be light in colour. The flower-buds are yellowish and less than, or about equal in length to, the green ovary. All six tepals are white, the inner three being broader than the outer three, and about twice as long; this means that, from above the flower, only the pointed tips of the outer tepals can be seen. This feature, and also the pale, off-white anthers, distinguish this species from *Libertia peregrinans*.

It is found on the edges of streams and among rocks: flowers in October and November.

Libertia peregrinans — Plate 2

LIBERTIA — **20**

A medium to tall perennial with green leaves, tending to become copper-coloured where they are exposed to sunlight. The flower buds are brownish and about equal in length to the ovary. All six tepals are white, the inner three being broader than the outer three, and less than twice as long. The inner tepals have a narrow base region so that, from above the flower, almost the whole of each outer tepal can be seen; this feature, and the dark orange-brown anthers and the copper-coloured leaves, separate this species from *Libertia ixioides*.

Locally common in the southern half of the North Island and in many areas of the South Island, it flowers in October to January.

*Schizostylis coccinea** — Plate 2

KAFFIR LILY — **50**

A tall, evergreen perennial, with a tuft of leaves up to 300 mm long, growing from a rhizome. The stems are stiff, up to 900 mm long, bearing spikes of distinctive and brilliant red flowers. The lower parts of the tepals are fused into a narrow tube about 25 mm long. The style is divided into three (sometimes four) fine, thread-like branches.

A garden escape, it lives in damp places beside roads, including drains: flowers from March to May.

Sisyrinchium iridifolium★ — Plate 2

SISYRINCHIUM — 20

A medium perennial with grass-like leaves about 5 mm wide. The tepals are all alike, being creamy white and striped with purple. On the insides of the tepals the purple colouring forms a six-pointed 'star'. Outside, the tepals are veined in purple, and are hairy on the lower half.

It lives in grassy places: flowers in December.

There are two other distinct forms of this plant growing wild in New Zealand, which may be subspecies. One is described as *Sisyrinchium 'blue'*★, and is a smaller plant with smaller (15 mm diameter) blue flowers with a yellow band and throat. The other is *Sisyrinchium 'yellow'*★, with even smaller (6 mm diameter) yellow flowers, and occurs only in the North Island.

Watsonia ardernei★ — Plate 3

WATSONIA — 35–45

A tall, robust perennial growing in clumps from corms about 40 mm in diameter. The leaves are broad and sword-shaped, with yellowish margins.

It differs from the other species of Watsonia because of its freely branching inflorescence of white flowers. The tube of the flower is a broad funnel. The style branches into three and each branch is shortly branched again into two.

It is a garden escape, living in grassy places: flowers in October and November.

Watsonia bulbillifera★ — Plate 3

WATSONIA — 30–40

A tall, robust perennial, growing in clumps from corms about 70 mm in diameter. It is distinguished from the other species of *Watsonia* by the numerous, small, reddish-brown, shining corms (or cormils) produced instead of flowers on the lower parts of the inflorescences. The inflorescence is less branched than that of *W. ardernei*, and often not branched at all. The flowers are brick red or salmon pink. As in the other species of Watsonia, the style branches into three and each branch is shortly branched again into two.

It is a very common garden escape, living on roadsides and in waste places: flowers in October and November.

Watsonia meriana★ **Plate 3**

WATSONIA **30–50**

A tall, robust perennial, growing in clumps from corms about 40–80 mm in diameter, it is a rather smaller plant than the two Watsonias described above. Like all Watsonias, it has three style branches, each subdivided again into two. The flowers are rose red, salmon pink, pinkish cream or (rarely) white. The plants with reddish flowers can be distinguished from *W. bulbillifera* by the lack of cormils; white-flowered plants are distinguished from *W. ardernei* by having unbranched or only slightly branched inflorescences.

A garden escape, it grows in grassy places: flowers in November and December.

Lily Family – Liliaceae

H ABfewW E – R
P 3+3 petal-like or sepal-like
M 3+3
F $\underline{(3)}$

The lilies are perennial herbs with bulbs. The petals and sepals are usually all petal-like, so that flowers appear to have six petals.

Many of the family are cultivated for their showy flowers, including the tulip, crocus, tiger lily, and fritillary.

Lilium tigrinum★ **Plate 1**

TIGER LILY **60–100 long**

A medium to tall perennial, growing from a large whitish or yellowish bulb. The flowering stem bears about five to six large flowers, with a characteristic appearance, having long protruding stamens and curved-back tepals. The tepals are bright orange-red in the most commonly found variety, but numerous cultivated varieties, such as the one illustrated in the photograph, have become naturalised.

It grows by roadsides and on waste areas of land: flowers in January and February.

Orchid Family – Orchidaceae

H A Si – I
P 3+3 see below
M 2 sometimes 1
F (3)

The members of this family are easily recognised by their distinctive flowers with left-right symmetry (zygomorphic). The pollen is usually produced in waxy masses, called pollinia; when a pollinating insect visits a flower, a whole pollinium is transferred to its body. The insect then visits another flower, where pollen is transferred to the stigma.

Insects usually visit flowers because they are attracted by the nectar or pollen, on which they feed. Some of the orchids have another way of attracting insect visitors: the bee orchids and some others of the orchid family produce scents that attract the insects sexually. In the bee orchids, parts of the flower resemble the abdomen of a bee. Male bees are attracted to, and attempt to mate with, the orchid. In doing so, some of the pollen from their bodies is transferred to the stigma.

The life cycle of orchids is closely dependent on fungi. When an orchid seed germinates, the seedling remains underground for two to four years. In this stage, called a protocorm, it relies for its food supply on fungi that live in its tissues. These fungi obtain food materials by breaking down dead plant and animal matter in the soil, and some of these materials are made available to the growing protocorm. At the end of this period the protocorm sends up leaf-bearing shoots above soil level. The plant then develops in the more usual way, manufacturing its own food materials by photosynthesis in sunlight. Even then, its dependence on fungi is not ended. The roots of the plants are covered in a sheath of fungus threads, a mycorrhiza. This grows out into the soil to absorb materials from the soil and to pass some of them on to the orchid. In return, the fungus receives photosynthesised materials from the orchid plant.

Many orchids live in forests, as epiphytes high on the branches of trees. They use the trees only for support, gaining the advantage of the brighter light conditions

MONOCOTS WITH LILY-LIKE FLOWERS

Allium triquetrum
p. 22

Bulbinella hookeri
p. 25

Bulbinella gibbsii
var. *balanifera* p. 24

Scilla non-scripta
p. 27

Asphodelus fistulosus
p. 24

Zantedeschia aethiopica
p. 23

Lilium tigrinum
p. 31

PLATE 2

IRISES

Aristea ecklonii
p. 28

Gladiolus undulatus
p. 28

Sisyrinchium iridifolium
p. 30

Schizostylis coccinea
p. 29

Libertia peregrinans
p. 29

Crocosmia ×crocosmiiflora
p. 28

Libertia ixioides
p. 29

Irises

Watsonia ardernei
p. 30

Watsonia bulbillifera
p. 30

Watsonia meriana
p. 31

Various monocots

Dianella nigra
p. 41

Phormium tenax
p. 41

Hedychium gardnerianum
p. 42

Tradescantia fluminensis
p. 26

PLATE 4

Orchids

Caladenia catenata
p. 34

Caladenia lyallii
p. 34

Thelymitra venosa
p. 40

Corybas trilobus
p. 36

Thelymitra longifolia
p. 39

Corybas macranthus
p. 36

Adenochilus gracilis
p. 33

Corybas rivularis
p. 36

Chiloglottis cornuta
p. 34

Orchids

Aporostylis bifolia
p. 33

Dendrobium cunninghamii
p. 37

Gastrodia cunninghamii
p. 38

Earina autumnalis
p. 37

Earina mucronata
p. 37

Microtis unifolia
p. 38

Orthoceras strictum
p. 38

Prasophyllum colensoi
p. 39

Pterostylis banksii
p. 39

PLATE 6

ICE PLANTS, AMARANTHS

Aptenia cordifolia
p. 48

Carpobrotus edulis
p. 48

Disphyma australe
p. 49

Tetragonia trigyna
p. 49

Amaranthus deflexus
p. 50

Alternanthera philoxeroides
p. 50

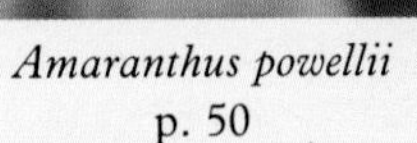

Amaranthus powellii
p. 50

CARROT FAMILY

Apium nodiflorum
p. 52

Apium prostratum var. *filiforme*
p. 53

Aciphylla colensoi
p. 51

Conium maculatum
p. 54

Anisotome haastii
p. 52

Daucus carota
p. 54

Foeniculum vulgare
p. 54

PLATE 8

DAISIES WITH WHITE-RAYED FLOWER-HEADS

Anthemis cotula
p. 58

Bellis perennis
p. 61

Leucanthemum vulgare
p. 59

Celmisia spectabilis
p. 63

Dolichoglottis scorzoneroides
p. 72

Celmisia verbascifolia
p. 64

Celmisia sessiliflora
p. 62

Celmisia semicordata
p. 62

Celmisia allanii
p. 62

available near the tops of the trees. Their roots absorb water from rainfall and nutrients from the debris that collects in crevices in the bark of the trees.

Orchids are some of the most showy of the native New Zealand herbs; indeed, the family as a whole is well known for its delicate, striking, and beautiful blooms. For this reason, orchids are widely cultivated as garden plants and house plants and are sold as cut flowers. They are customarily worn for special occasions, such as weddings and other celebrations. One species of tropical climbing orchid, *Vanilla*, is the source of vanilla food flavouring, which is obtained from its seeds.

Adenochilus gracilis — Plate 4

ADENOCHILUS ORCHID — 10–15

A low to short perennial with a single, oval, green leaf, about 10–30 mm long, usually about half-way up the stem. There are also two to three short, narrow bracts. The lateral tepals are alike in shape and greenish white in colour. The central sepal is deeply hooded, covering the column and tongue. The tongue is short and broad, spotted or striped with red, and with a distinctively sharp-pointed, curved tip. The tongue bears a central row of yellowish calli which extends on to the tip.

It is common in light shade: flowers from May to September.

Aporostylis bifolia — Plate 5

ODD-LEAVED ORCHID — 15–25

A short, hairy perennial, reproducing by tubers. At the bottom of the stem are two softly hairy leaves; the lower leaf is longer and broader than the upper one, and both are blotched with brown. At the top of the single stem is a white, or sometimes pink, flower. The petals have a pink stripe down the middle. The middle petal is wider than the others, arched over to form a shallow hood. The tongue is broad and oval, with two rows of yellow calli near the base. The tongue has fallen from the flower on the left in the photograph, revealing the column; this has two narrow wings running up either side of it, becoming broader and forming two lobes, one on each side of the anther.

It lives in wet areas, such as bogs and hollows in grassland and scrub: flowers in December and January.

Caladenia catenata — Plate 4

CALADENIA ORCHID — 10–20

A short, slightly hairy perennial, with a single, narrow (up to 4 mm wide) leaf, which has both short, glandular hairs and long, non-glandular hairs. The other native species, *Caladenia lyallii*, has broader, slightly hairy leaves. Flowers are few on a stem. The tongue has three lobes; the middle lobe is triangular, slightly curved back (compare *C. lyallii*), with yellowish calli on its edges; the two broad, side lobes are turned upward, with reddish stripes. There are two (compare *C. lyallii*) rows of red, yellow-tipped calli on the central area of the tongue.

It grows in shady places in scrubby areas: flowers in September and October, earlier than *C. lyallii*.

Caladenia lyallii — Plate 4

WHITE FINGERS — 15–25

A short, slightly hairy perennial, with a single leaf up to 8 mm wide, which is only slightly hairy. Flowers are white or pink. The tongue has three lobes: the middle lobe is triangular and curled tightly back; the two broad side lobes are turned upward, with reddish stripes. The four rows of yellow calli, on the central area of the tongue, distinguish this species from *C. catenata*, as well as the broader, only slightly hairy leaves.

It lives in beech forest, scrub and grassland: flowers from November to February, later than *C. catenata*.

Chiloglottis cornuta — Plate 4

CHILOGLOTTIS ORCHID — 15

A short, hairless perennial, reproducing by tubers. Each plant has one green and rather fleshy flower. The two leaves are close together at the base of the stem; they are hairless, upright, spreading slightly, with the flower between them. There is a green, fleshy bract close below the flower, its wide base sheathing the stem. This may be mistaken for a third leaf, especially when the plant is fruiting and the stem between the bract and leaf becomes longer.

It lives in shady places, usually damp ones: flowers from October to February.

Understanding orchid flowers

Like other monocots, orchid flowers have six tepals, arranged in two whorls or 'rings'. The outer whorl has three sepals and these alternate in position with the three petals of the inner whorl.

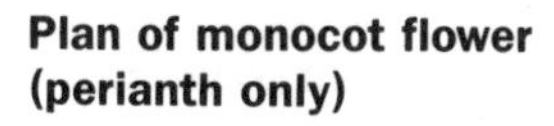

Plan of monocot flower (perianth only)

Plan of orchid flower (perianth only)

In orchids, the tepals are all petal-like in shape and texture, though they may be green in colour. The ovary of orchids is inferior and, in most of them, the ovary or the flower stalk is twisted; this rotates the flower by half a turn (180°). Thus an orchid flower is usually 'upside down'. An exception is *Prasophyllum*.

In some orchids (for example, *Thelymitra*), the tepals are all very similar in size, shape and colour. In others, one or more of the tepals may be different from the others. Frequently, the middle sepal is much larger and shaped to form a hood over the rest of the flower. As well as this, the middle petal becomes expanded to form a prominent 'tongue' (labellum). These drawings show the flower plan and a vertical half-section of a typical orchid flower.

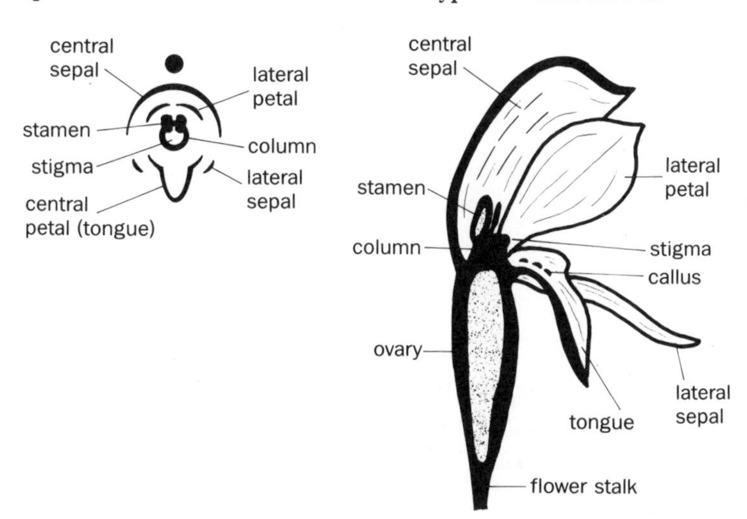

The shape of the tongue is often an important feature for identifying orchids. The tongue may have hard swellings on it, known as calli.

The centre of the orchid flower is occupied by the column, which carries the stamens and stigmas. The stamens have no filaments, the anthers being mounted directly on the column.

Corybas macranthus — Plate 4

SPIDER ORCHID — 10

Eight species of *Corybas* are native to New Zealand. The three species illustrated here are all low-growing perennials with a single green leaf, which is broad and flat. They have a single flower on a short stalk, so that it is below the leaf. The lateral sepals and petals are very narrow and elongated, giving the name 'spider orchid' to this genus. A distinctive feature of this species is its oval, fleshy leaf about 35 mm long on a relatively long stalk. The apex of the leaf is rounded with a minute, pointed tip. The leaf is silvery below. The flower is coloured dark red and white with prominent 'spider legs'. The central sepal is narrower than in *C. trilobus* and its tip is sharply pointed. The tongue is longer than the central sepal, very broad, and curved back, and its edge has many small teeth.

It lives in damp shady places, such as the forest floor, stream banks, and shaded grassland: flowers from October to December.

Corybas rivularis — Plate 4

SPIDER ORCHID — 10

The leaf of this species has no stalk and is distinctly longer than broad. It is heart-shaped at the base and tapers gradually to a sharply angled apex. The edges of the leaf are wavy. The central sepal is much longer than in the two species above, tapering to a long, narrow point.

It lives on the forest floor, flowering from September to November.

Corybas trilobus — Plate 4

SPIDER ORCHID — 10

The leaf of *Corybas trilobus* is rounded, about 20 mm in diameter and on a relatively long stalk. The tip of the leaf is sharply pointed, with rounded lobes on either side of it, as seen in the photograph. The flower is red and green with prominent white 'spider legs'. The central sepal is arched over to form a deep hood; its tip is blunt. The tongue, which is curved back, is smaller than the hood and enclosed within it, and its edge has many small teeth.

The plant is found growing in damp shady places, such as the forest floor, and under scrub: flowers from June to December.

Dendrobium cunninghamii — Plate 5

LADY'S SLIPPER ORCHID — **20–25**

A short perennial with shiny yellow, cane-like stems up to 1.5 m long, with many drooping branches. The leaves are well spaced out along the stem and are narrow (30–50 mm long, 3 mm wide). The flowers are larger than most orchids, and occur singly or in clusters of up to six. The tongue has three lobes: the central lobe is broad and white; the two lateral lobes are coloured red and stand up. The base of the tongue has four to five yellowish ridges on it.

It grows as an epiphyte on trees in well-lit areas of forest, also on fallen logs, and on rocks: flowers from December to February.

This is the southernmost member of the very large and showy *Dendrobium* genus. The name *Dendrobium* comes from the Greek *dendron* (a tree) and *bios* (life); this refers to its habit of living on trees.

Earina autumnalis — Plate 5

RAUPEKA – EASTER ORCHID — **13**

A medium perennial with short, erect stems, drooping if the plant grows longer (up to 1 m), but usually unbranched. The leaves are 40–120 mm long, 5–8 mm wide, densely covering the stem. The showy inflorescence of up to 40 waxy, white, fragrant flowers distinguishes this species from *E. mucronata*. The sepals are alike in being triangular in shape, about the same length as the more rounded lateral petals. The base of the tongue is yellow and its sides curve up to partly enclose the column.

This plant is epiphytic on trees, but also grows on rocks and banks. It flowers from February to April, so is often in bloom at Easter.

Earina mucronata — Plate 5

PEKA-A-WAKA – BAMBOO ORCHID — **10–12**

A medium perennial with short, erect stems, drooping if the plant grows longer (up to 1 m), but usually unbranched. The leaves are 60–150 mm long, 4–6 mm wide, densely covering the stem. The leaf sheaths have small, dark spots. The inflorescence is a small, drooping spray of up to 10 greenish cream flowers. The sepals and lateral petals are more or less alike in size, shape and colour. The tongue is broader and longer than that of *E. autumnalis*, and is yellowish in colour; its tip is expanded into a broad lobe that projects from the flower, another feature distinguishing this species from *E. autumnalis*.

It grows as an epiphyte on trees, as well as on rocks and banks: flowers from September to November.

Gastrodia cunninghamii — Plate 5

HUPEREI – PEREI – MAUKUUKU – BLACK ORCHID — 14 long

A tall, leafless plant with brown stems, bearing a spike of purplish brown flowers. The flowers are tubular with five lobes. The tongue is enclosed in the tube and has wavy edges. The column is much shorter than the tongue. The flowers are mottled with lighter swollen spots, a feature that distinguishes this from other species of *Gastrodia*; the inside of the flowers is creamish white.

It is found in lowland forests: flowers from November to February.

A related plant is *Gastrodia sesamoides*, found in open forest and scrub north of latitude 42°S. In this plant, the column is almost the same length as the tongue. Both plants obtain nourishment by associating with soil fungi that live as parasites on the roots of forest trees. The roots were eaten by the Maori, for use as food (perei) in winter.

Microtis unifolia — Plate 5

ONION ORCHID — 3 long

A tall, hairless perennial with a single, rolled leaf, often reaching higher than the spike of flowers. The raceme consists of many small, densely packed flowers with green tepals. The uppermost sepal (the dorsal sepal) is formed into a hood (compare with *Prasophyllum*). The 'tongue' is broad, roughly rectangular, with an irregular outline and a notch in the tip.

Widespread in open places, it flowers from October to February.

Orthoceras strictum — Plate 5

IKAIKA — 10

A tall, perennial plant growing from a tuber. There are several grass-like leaves on each stem. The stalk of the inflorescence has a slight zig-zag, as can be seen in the photograph. The long (15–30 mm) lateral sepals help to identify this plant; the flower diameter quoted above excludes these sepals. Another distinguishing feature is the single yellow callus at the base of the tongue, visible in the lower flower of the photograph.

It grows in open scrub and on banks: flowers from December to March.

Prasophyllum colensoi **Plate 5**

LEEK ORCHID **7 long**

A short, hairless perennial with a single, long, rolled leaf, often reaching higher than the raceme of flowers. The raceme consists of many small, fleshy, reddish green or yellowish green flowers. The flower is upside down compared with most other orchids (see p. 35). Instead of a single central sepal above the flower, there are two partly fused lateral sepals (compare with *Microtis*). The central sepal is below the flower and is slightly curved, but not hooded. Another result of the flower being upside down is that the tongue is above the column instead of below it. The tongue tapers to a narrow tip, with a single callus extending almost as far as the tip.

Widespread in grassland, especially where it is damp or boggy, it flowers from November to February.

There are three other species of *Prasophyllum* that are reasonably common in New Zealand. *P. patens*, found only in the North Island, has the long rolled leaf of *P. colensoi*, but the tongue is wider and has a wavy tip, with no callus at the tip. The other two species have only a short leaf, not reaching above the raceme. The flowers of *P. pumilum* have a greenish tongue, while the tongue of *P. nudum* is red and is fringed with fine hairs. *P. pumilum* is confined to the North Island.

Pterostylis banksii **Plate 5**

TUTUKIWI – COMMON GREENHOOD ORCHID **40**

A low to short perennial, with four to six grass-like, keeled, leaves on each plant. The leaves are up to 250 mm long and slightly sheathe the stem. Each plant has a single flower, the central sepal being green, helmet-shaped and usually more than 25 mm tall. The dark green veins of the central sepal are a distinctive feature. Another feature is that the lateral sepals project sideways from the flower, and have an orangy pink, tail-like appendage, longer than the sepal itself.

It lives on the floor of forests: flowers from October to December.

There are 19 species of *Pterostylis* in New Zealand, of which *P. banksii* is the commonest.

Thelymitra longifolia **Plate 4**

MAIKUKU – COMMON SUN ORCHID **8–15 long**

A low to medium, hairless perennial. The stem bears a single, strap-shaped leaf,

10–20 mm wide. The inflorescence has up to 10 (sometimes more) flowers with white tepals, usually reddish green on the outside; the flowers open widely in the sun. The flowers do not look like a 'typical' orchid as the six tepals are all similar in shape and size.

It grows on rocks, banks and open ground, and is probably the commonest wild orchid: flowers from October to February.

The tubers were eaten by the Maori, either raw or boiled.

Thelymitra venosa — Plate 4

STRIPED SUN ORCHID — **10–15 long**

A short to medium, hairless perennial. As in *T. longifolia*, the stem bears a single leaf, but the leaf of this species is narrower and thickened at the edges and midrib. The blue flowers, usually striped with a darker purplish blue, distinguish this species from *T. longifolia*.

It lives in wet, poorly drained areas: flowers from December to March.

Other sun orchids with blue tepals include *T. formosa*, with blue tepals without stripes or spots; *T. pulchella*, with pinky-blue or white tepals with blue stripes (though it may also have plain pink tepals); *T. decora*, with lavender-blue tepals with dark spots, mainly on the petals; and *T. ixioides*, with blue tepals with darker blue spots.

Harakeke Family — Phormiaceae

HS Eq Si – R
P 3+3 outer tepals may be smaller, or differently coloured
M 3+3
F (3)

The leaves of this family are at or near the base of the stem; they are alternate in two opposite rows, their bases folded and sheathing each other (equitant, see inside back cover). Plants may be massive and shrubby, but their stems do not increase in thickness over the years as do the trunks of a truly woody plant. There is a distinct 'joint' at the base of each flower, more clearly seen when the flowers have dropped off.

In economic terms, the most important family member is New Zealand flax, cultivated for its fibrous leaves. It is also grown in Central Africa, the USA and Mauritius. A large species of *Dianella* is cultivated for the same reason in Hawaii.

Dianella nigra — Plate 3

TURUTU – BLUE-BERRY – INK-BERRY — 8

A medium, perennial plant with fans of long, narrow, glossy leaves arising from short, woody shoots. The leaves are edged dark reddish brown at the base, and they are strongly keeled and folded. Just above the region where it sheathes a younger leaf, the two sides of each leaf are pressed tightly together, almost becoming fused. Above, the leaf opens out to be more or less flat, though still with a noticeable keel. The inflorescence is a much-branched, wiry stalk bearing small, greenish white flowers and greenish, oval, fleshy fruits. The six tepals are alike in colour and size, the inner three being slightly wider than the outer three. The six stamens are distinctive, having swollen, yellow filaments, as seen in the photograph. Later the fruits ripen to develop a greyish white to strong blue colour.

It is found on banks and on banked roadsides, often in shady situations, and on the forest floor, especially at the edges of tracks: flowers in November and December.

Phormium tenax — Plate 3

HARAKEKE – NEW ZEALAND FLAX — 25–50 long

Tall, perennial, tufted plants, with long (up to 3 m) stiff, pointed leaves. The inflorescence is up to 5 m or even 6 m long, consisting of 50 or more dull red flowers; it produces dark-coloured, straight pods that are three-sided and erect when ripe.

The related species, wharariki or mountain flax (*Phormium cookianum*), is similar but has yellowish flowers. Its ripe pods are rounded in section, hang downward, and are spirally twisted. It lives in various localities from coastal cliffs to high country; it may occur at the same sites as *P. tenax* and may hybridise with it.

This plant lives mainly in lowland swampy areas, and flowers from November to December.

Both species of flax are visited by bellbirds, tui and starlings, who feed on the nectar. The nectar was collected by the Maori to use as a drink and for sweetening. Fibres were extracted from the leaves and used for making textiles and cord. Sir Joseph Banks, on his visit to New Zealand in 1769 with Captain Cook, was impressed by the quality of the fibre. He wrote that it is 'of a strength superior to hemp. . .shining almost as silk and surprisingly strong'. The name *Phormium* comes from the Greek *phormion*, 'a mat', and refers to one of the practical uses of the leaf fibres.

Ginger Family – Zingiberaceae

H A E – I
K (3) fused at the base to form a narrow tube
C (3) fused at the base to form a narrow tube
M 3+3 one fertile and several petal-like infertile staminodes
F (3)

The plants are perennials, sending up usually unbranched, leafy stems from fleshy root-stocks. The flowers are large and complicated in structure.

The family includes several plants that are of importance in the kitchen. Ginger is used in flavouring, either as the dead rhizome, whole or ground, and comes from *Zingiber officinale*, a native of South-East Asia. Other members of the family produce turmeric and cardamoms.

Hedychium gardnerianum⋆ — Plate 3

WILD GINGER — **70 long**

Hedychium is the only genus of this family growing wild in New Zealand. It is a typical member of the family and conforms to the family description given above. The leaves are large and have no leaf-stalk or, at most, a very short one; they are arranged in two rows on opposite sides of the stem. The pale lemon flowers with the prominent red filament of the stamen distinguish this from other species.

It grows on roadsides, and flowers in February and March.

The other wild ginger in New Zealand is *H. flavescens*⋆. This has a similar appearance to *H. gardnerianum*, but its inflorescence is more cone-shaped, with creamish white, fragrant flowers, and a creamish white stamen filament. Several species of *Hedychium* are cultivated in gardens for ornament and may occasionally escape to roadsides and similar areas.

DICOTS

The main differences between this class and the other main class of flowering plants, the Monocot Class, are described on p. 17. The Dicot Class is the larger of the two classes and includes a wider range of types. Many of the trees and shrubs belong to this class. Most of the families of dicots are represented in New Zealand; some are represented by only one or two genera, while others are represented by many genera.

Well represented families include:

- Daisy Family – Asteraceae
- Mustard Family – Brassicaceae
- Pink Family – Caryophyllaceae
- Pea Family – Fabaceae
- Mint Family – Lamiaceae
- Buttercup Family – Ranunculaceae
- Foxglove Family – Scrophulariaceae

If you concentrate first of all on getting to know the main features and the commonest members of these families, you will soon be able to identify, or at least place in their families, the majority of wild flowers you find in the field.

IDENTIFYING DICOTS

Since there are so many families of dicots, and the essential differences between them are not easy to decide in the field, we use a feature table for identifying them: you will find this quicker than running through a long key. The table is particularly useful if the information you have about a plant is limited. Another advantage is that it will tell you the family of a plant (provided the family is in this book, which it probably is) even though its genus or species is not described here.

The feature table below lists all the 53 dicot families in this book. The columns of the table refer to nine features of plants that are useful in deciding which family they belong to. A 'y' in any column tells you that all (or almost all) members of the family possess that feature. For example, members of the Apocynaceae (Periwinkle Family) all have opposite leaves and united petals.

An 'n' in any column means that no (or very few) members of the family possess that feature. In that event they will show a feature that is the opposite in some way.

The features are grouped in nine contrasting pairs:

If 'y' it has:	*If 'n' it has:*
opposite leaves (includes whorled)	alternate leaves
compound leaves	entire or simple leaves
stipules present	stipules absent
flowers irregular (zygomorphic)	flowers regular
sepals united	sepals free (or no sepals)
petals united	petals free (or no petals)
stamens many, no definite number	stamens few, definite number
ovary inferior	ovary superior or partly superior
carpels free or only one	carpels united

If you do not know the meanings of any of the words used above, refer to the inside of the back cover. With only a little experience it soon becomes easy to decide whether a given plant specimen is to be rated 'y' or 'n' for each of the nine features.

In some families there may be some species which rate 'y' for a given feature, but there are several or many other species which rate 'n' for that feature. In such cases the table shows 'b' (for 'both'). For example, in the Amaranth Family some species have opposite leaves and others have alternate leaves.

Take a strip of paper one or two centimetres wide and lay it across the table, just below the column headings. Look at your plant and, for each feature, decide if it rates as 'y' or 'n'. Write 'y' or 'n' on your slip just below the corresponding heading. Sometimes you will not be able to make a decision: for example, it may be hard to decide if the ovary is inferior or superior, or the stipules may drop off soon after the leaves have opened. If you are uncertain whether to put 'y' or 'n', leave a blank – try not to guess!

Now lay the strip just below the first line (Aizoaceae) of the table. Do all the 'y's and 'n's match? If a feature has a 'b' in the table, ignore it, as it is no help in identification for this family. Similarly, if your strip has a blank, ignore this too. Apart from the features which must be ignored, *all* other 'y's or 'n's in the first row of the table must match with the 'y's and 'n's on the strip. If you get a perfect match (ignoring 'b's and blanks), make a note of the family name given on the left. If the strip and the row fail to match in any one or more column, your plant cannot be a member of that family.

When you find a row that completely matches your strip, look at the column on the right. This lists certain other features shared by all or most members of the family and may help you confirm or reject the family identification.

Run down all of the rows in the table, noting the names of possible families. Sometimes there are two or possibly more matching families, but often there will be only one. The page numbers on the right tell you where the descriptions of plants of that family begin. Turn to those pages to identify the genus or species.

This system of identifying plants is easily made into a computer program. At the end of the book we give an outline program showing how to put these tables on to your home computer and thus speed up your identification of plants.

Dicot features table

Family	Leaves opposite	Leaves compound	Stipules present	Flowers irregular	Sepals united	Petals united	Stamens many	Ovary inferior	Carpels free	Confirming features	Page
Aizoaceae	y	n	n	n	n	n	y	y	n	Fleshy leaves; many petals and sepals	48
Amaranthaceae	b	n	n	n	n	n	n	n	n	Small chaffy fls; T3–5; M5 opposite C	50
Apiaceae	n	y	b	n	n	n	n	y	n	Umbels; aromatic; K5; C5; M5	51
Apocynaceae	y	n	n	n	n	y	n	n	n	Twining; long pods; M5	55
Asteraceae	b	b	n	b	n	y	n	y	n	Fl-heads; M5 fused; K is pappus	56
Boraginaceae	n	n	n	n	y	y	n	n	n	Bristly-hairy; fl stem curled; 4 nutlets	85
Brassicaceae	n	b	n	n	n	n	n	n	n	Cross-flowers; M 4-long + 2-short OR 4	87
Campanulaceae	n	n	n	n	y	y	n	y	n	Usually blue or violet; fls bell-shaped	93
Caprifoliaceae	y	n	n	b	y	y	n	y	n	Shrubs; woody climbers	94
Caryophyllaceae	y	n	b	n	b	n	n	n	n	Stem nodes swollen; M twice C	95
Chenopodiaceae	n	n	n	n	y	n	n	n	n	Inconspicuous fls; not chaffy; pl mealy	99
Clusiaceae	y	n	n	n	n	n	y	n	n	Stamen bundles; dotted lvs & K; yellow fls	101
Convolvulaceae	n	n	n	n	n	y	n	n	n	Twining stems; funnel shaped fls; latex	102
Crassulaceae	y	n	n	n	b	b	n	n	y	Fleshy; F same no as C	104
Dipsacaceae	y	b	n	y	y	y	n	y	n	Fl-heads; M4	105
Droseraceae	n	n	n	n	y	n	n	n	n	Long glandular hairs; insectivores	106
Ericaceae	b	n	n	n	y	y	n	n	n	Low shrubs; M twice C	107
Euphorbiaceae	b	n	y	n	n	n	n	n	n	Latex; fls unisexual; C=0	108
Fabaceae	n	y	y	y	y	n	n	n	n	'Pea' fls; pods	109
Fumariaceae	n	y	n	y	n	n	n	n	n	Stems brittle; petal spurs	117
Gentianaceae	y	n	n	n	y	y	n	n	n	Sessile lvs; bell-shaped fls; hairless	118

Family	Leaves opposite	Leaves compound	Stipules present	Flowers irregular	Sepals united	Petals united	Stamens many	Ovary inferior	Carpels free	Confirming features	Page
Geraniaceae	b	n	y	b	b	n	n	n	n	'Storksbill' fruits	120
Goodeniaceae	n	n	n	y	y	y	n	y	n	Pollen cup around stigma; C tube split	124
Haloragaceae	b	n	n	n	y	n	n	y	n	Often aquatic; fruits hard and squarish	124
Hectorellaceae	n	n	n	n	n	n	n	n	n	Cushion-pl; K2; C5; M5; (Hectorella)	125
Lamiaceae	y	n	n	y	y	y	n	n	n	Square stems; aromatic; 4 nutlets	126
Linaceae	b	n	n	n	n	n	n	n	n	Unbranched erect stems; small narrow lvs	129
Lobeliaceae	n	n	n	y	y	y	n	y	n	M joined in a tube; C tube split	130
Lythraceae	y	n	n	n	y	n	n	n	n	Epicalyx; calyx tube	132
Malvaceae	n	n	y	n	y	n	y	n	n	Lobed leaves; M filaments joined	133
Onagraceae	b	n	n	n	y	n	n	y	n	Calyx tube; M4 or 8; K4; C4	134
Orobanchaceae	n	n	n	y	y	y	n	n	n	No chlorophyll; parasitic; scale lvs	137
Oxalidaceae	n	y	y	n	n	n	n	n	n	Trefoil leaves; orange calli on lvs & spls	137
Papaveraceae	n	b	n	n	b	n	y	n	n	K2; falling off in bud; latex	139
Passifloraceae	n	n	y	n	n	n	n	n	n	Corona; tendrils	140
Phytolaccaceae	n	n	n	n	y	n	n	n	y	Small fls in dense spikes; C=0	141
Plantaginaceae	n	n	n	n	y	y	n	n	n	Basal lvs; parallel vns; small chaffy fls	142
Polemoniaceae	b	b	n	n	y	y	n	n	n	K5; C5; M5	143
Polygonaceae	n	n	y	n	n	n	n	n	n	Ochreae; small fls; swollen stem nodes	143
Portulacaceae	b	n	n	n	b	n	n	n	n	Fleshy leaves; satiny petals; M3+; K irreg	147
Primulaceae	y	n	n	n	y	y	n	n	n	Leaves basal; M opposite C	148
Ranunculaceae	b	b	n	n	n	n	y	n	y	P sometimes of 1 whorl; M in spiral	150

Family	Leaves opposite	Leaves compound	Stipules present	Flowers irregular	Sepals united	Petals united	Stamens many	Ovary inferior	Carpels free	Confirming features	Page
Resedaceae	n	n	n	y	n	n	y	n	n	K4–8; C4–8; M3–40	155
Rosaceae	n	y	y	n	n	n	y	n	y	Epicalyx; M & F many (except *Acaena*)	156
Rubiaceae	y	n	y	n	n	y	n	y	n	Leaves appear whorled	160
Scrophulariaceae	b	n	n	n	y	y	n	n	n	M fewer than C; F2	161
Solanaceae	n	b	n	n	y	y	n	n	n	'Spire' of fused anthers; nasty smell	168
Stylidiaceae	n	n	n	n	y	y	n	y	n	M2 united into column	170
Tropaeolaceae	n	b	n	y	n	n	n	n	n	Lvs peltate; sepal spur; clawed petals	171
Urticaceae	b	n	y	n	n	n	n	n	n	Stinging hairs; unisexual fls	172
Valerianaceae	y	n	n	y	n	y	n	y	n	Dichotomous branching; M1–4; petal spur	173
Verbenaceae	y	n	n	y	y	y	n	n	n	Square stems; thorny; narrow C tube	173
Violaceae	n	n	y	y	n	n	n	n	n	Petal spurs; M fused in ring; solitary fls	175

Mesembryanthemum Family – Aizoaceae

HS OfewA E – R
K 3–8 usually unequal
C many rarely none
M many rarely one or few
F (2–5)

Plants in this family are usually very succulent, with thick, fleshy leaves. Leaves are often rounded or triangular in section, and may be covered with tiny, rounded swellings, giving the leaf a shiny, glittering appearance.

One plant of economic importance is New Zealand spinach.

Aptenia cordifolia★ Plate 6

APTENIA 20–25

A low, spreading, perennial herb with stems up to 300 mm or more long. The leaves are fleshy but flat, and oval to triangular in shape. The tiny swellings on the leaf, seen in the leaf at bottom centre of the photograph, are clear but not glittering, and also occur on the sepals. The four sepals, all larger than the petals, confirm the identity of this species.

It grows in rocky or sandy areas on the coast: flowers all the year round.

The genus name *Aptenia* means 'without wings', referring to the lack of wings on the fruit. The heart-shaped leaves give it the species name *cordifolia*.

Carpobrotus edulis★ Plate 6

ICE PLANT – CAPE FIG – HOTTENTOT FIG 80–100

A short, spreading perennial with long, woody stems, up to 6 m long. The leaves are thick and fleshy, sharply triangular in section, and have a bitter taste. The yellow petals fade to a pinkish orange colour.

It is found on the coast, growing on cliffs and sand dunes: flowers from October to February.

The inner part of the fruits may be eaten, giving the plant its scientific name *Carpobrotus*, which comes from Greek and means 'edible fruit'. The leaves too are edible, usually being pickled first to preserve them. This plant has also been known as *Mesembryanthemum*, from which is derived one of its common names 'Sally-my-handsome', popular in Cornwall, England.

Disphyma australe — Plate 6

HOROKAKA – NGARANGARA – MAORI ICE PLANT — 20–40

A short, spreading perennial with thick, fleshy leaves. The leaves are triangular in section but more rounded than those of *Carpobrotus*, and end in a short, sharp point. The flowers are solitary, with white or pink petals.

Widespread in coastal regions, it flowers from October to January.

The inner part of the fruit and the pickled leaves can be eaten. A hybrid genus, known as ×*Carpophyma*, is also found; this is a naturally occurring cross between *Carpobrotus edulis* and *Disphyma australe*. The flowers are of intermediate size (45–60 mm in diameter) and are of various colours including pink, yellow, orangy pink and white. No fruits are formed by this hybrid.

Tetragonia trigyna — Plate 6

BEACH SPINACH — 7

A short, spreading herb with fleshy leaves. Leaves are oval, rhomboid or triangular in shape. They are alternate, unlike most members of this family. The flowers have no petals, the perianth consisting of four small, yellowish sepals. They are usually solitary but may occur in pairs.

It is found in coastal areas, on sand dunes and beaches: flowers from November to March.

The leaves of young plants are cooked and eaten. A related plant is kokihi, New Zealand spinach (*Tetragonia tetragonioides*).

Amaranth Family — Amaranthaceae

HfewST O Si – R
T 3–5 small, chaffy
M 5 sometimes fewer, opposite the tepals
F 1
The long inflorescences of many small flowers, such as in *Amaranthus deflexus* (below) and in the popular garden plant love-lies-bleeding, are typical of this family.

Various species of *Amaranthus* are cultivated as amaranthus spinach.

*Alternanthera philoxeroides** — Plate 6

ALLIGATOR WEED — 10–15

A spreading water-plant with hollow stems up to 2 m long, floating on the water. The ends of the stems turn upward and bear rounded inflorescences about 10–20 mm in diameter.

Common in streams, swamps and in other slowly moving or stagnant water in north Auckland, it also occurs on banks and may spread to surrounding areas of damp soil.

*Amaranthus deflexus** — Plate 6

PROSTRATE AMARANTH — 2–3

A low, mat-forming, annual herb. The leaves are dark green, and the flowers, with two to three tepals, are only a little longer than the bracts. The inflorescence is a dense spike of greenish or pinkish flowers.

It is common on waste land, especially in gravelly areas: flowers from November to May.

*Amaranthus powellii** — Plate 6

REDROOT — 2–3

A low to medium, hairy, annual herb, with angular stems, often red. Flowers with four or five green tepals, unequal in length, tapering to a sharply pointed tip. The inflorescence has a main spike up to 250 mm long, with a few leaves toward the lower end. There are also a few shorter spikes in the axils of the upper leaves, as seen in the photograph. It is often confused with another redroot, *A. retroflexus**, the tepals of which have square-cut or rounded tips, and which has its shorter spikes branching from the main spike.

It lives in and around town areas: flowers from December to February.

Carrot Family – Apiaceae

H A CfewESi – R
K 5 **or** (5) usually very small
C 5 usually white, sometimes yellow
M 5
F (2)

This family is named after a typical genus *Apium*, which includes wild celery. The family is also known as the Umbelliferae because of the radiating umbrella-like stalks of its inflorescence. These umbels may be simple or compound. The leaves are usually large and pinnate. The pinnae may be pinnate themselves (two-pinnate). In some species the leaves are divided pinnately yet again (three-pinnate) or even again (four-pinnate). The presence or absence of bracts (leaf-like scales) at the bases of the main stalks of the umbels is often used as a feature for recognition; there may also be smaller bracts (bracteoles) at the bases of the stalks of the sub-umbels. In some species the bracts are large and conspicuous.

Many members of this family are grown for use in the kitchen:

- Swollen roots eaten: carrot, parsnip, celeriac.
- Leaves used for flavouring: parsley, fennel, dill, chervil, sweet Cicely, lovage.
- Leaf-stalks eaten or used for flavouring: celery (but in Indonesia the leaf *blades* are eaten and the leaf stalks thrown away!), Florence fennel.
- Young stalks and leaf-stems crystallised with sugar: angelica.
- Fruits used for flavouring: caraway, cumin, coriander, anise, celery.
- Seeds of angelica are used as flavouring in vermouth and in chartreuse liqueurs.

Aciphylla colensoi — **Plate 7**

WILD SPANIARD — 5

A tall, robust perennial forming large tufts. The leaves are about 300–500 mm long, with two to four pairs of leaflets, having strongly toothed margins and ending in sharp spines. Each leaf has two stipules, also ending in spines. The midribs of the leaflets are stout, and coloured reddish, orange or yellow. The inflorescence is up to 2.5 m long, with scented flowers, in small, dense umbels. The lower bracts of the inflorescence taper to sharp spines.

It grows in the South Island, in grassland, herbfield and snow tussock herbfield, from 900 m to 1500 m: flowers from November to February.

A spear grass of similar appearance is *A. scott-thomsonii*; this is the tallest spear grass, with flower stems reaching up to 3 m. It is larger overall than *A. colensoi* and differs in having less prominent, green to pale yellow midribs.

Anisotome haastii — Plate 7

PINAKITERE — 7–9

A medium, robust perennial with leaves two-pinnate, three-pinnate, or occasionally four-pinnate, 150–250 mm long and 60–120 mm wide. Leaf segments are narrow and end in hair-like tips. The leaves are 'flat', the leaf segments lying all in one plane, and the stalks of the leaves and bracts have prominent sheathing bases. The inflorescence consists of many compound umbels, 50–80 mm in diameter, with bracts ending in hair-like tips.

It grows in scrub and snow tussock herbfield, from 600 m to 1500 m: flowers from October to February.

There are 16 species of *Anisotome* native to New Zealand, possibly more if some of the forms presently recognised as varieties are in fact separate species. The species illustrated here is representative of many of the genus.

*Apium nodiflorum** — Plate 7

FOOL'S WATERCRESS – WATER CELERY — 1–2

A short, spreading perennial, with rooting stems. The one-pinnate leaves at first glance resemble those of watercress, but the margins of watercress leaves are smooth, while those of fool's watercress are toothed. As can be seen in the photograph, the terminal leaflet is often lobed. The umbels are unstalked or have very short stalks and are located opposite the leaves; they are 20–40 mm in diameter. Bracteoles are present, oval to triangular in shape, and about four to eight to each sub-umbel.

It lives in aquatic habitats, such as swamps and the edges of streams. Locally common in the North Island, it probably occurs also in the South Island, but has only been recorded there once. It flowers from November to February.

The spaniards and spear grasses

The genus *Aciphylla* (meaning 'sharp leaf') has 40 or more species, 39 of them native to New Zealand. They are found from sea level to 1850 m, but are usually commonest in subalpine and alpine regions; they are recognised as a group by their pinnate leaves, which arise from the base of the plant and often have a 'jointed' appearance. There is a sharp spine at the end of each pinna, and there are also spines on the stipules and bracts: these spines can cause painful injuries to those who accidentally come into forcible contact with the plant. Many of the genus are large, up to 3 m tall, with leaves up to 1 m long. In the larger species, the inflorescence consists of compound umbels on a long, stout central stalk, a metre or more long, or on a branching stalk arising from the main stem. The sepals are very small and poorly developed. The petals are small, and coloured white or yellow. The species described on p. 51 is typical of the larger spaniards.

Apium prostratum var. *filiforme* — Plate 7

TUTAEKOAU – MAORI CELERY — 1–2

A short, spreading to upright perennial, its stems do not take root. Leaves are one-pinnate, the segments being divided into rounded lobes, and the lower leaves often consist of only three segments. In the variety *prostratum*, the segments of the leaflet are longer and narrower. In the variety *denticulatum* the leaflets are more divided and the segments have finely toothed edges. In all varieties, the umbels are unstalked or with very short stalks, located opposite the leaves, as in *Apium nodiflorum* above, but there are no bracts or bracteoles.

It is found in coastal areas, growing on rocks, gravel or mud.

This is one of the two common plants collected in quantity by Captain Cook in 1796, as a preventative of scurvy; in those days scurvy was a frequent disease of seamen, officers and crew alike. Unless precautions were taken, many men would die of scurvy during a voyage lasting several months, because the food carried on board lacked vitamin C, without which the fatal disease eventually develops. Fresh vegetables such tutaekoau were an essential and rich source of the vitamin.

Leaves and stems may be eaten; the seeds may be used as flavouring. Make sure that you have identified the plant correctly, as it is possible to confuse it with other members of this family, such as the poisonous hemlock, *Conium maculatum*. The related species *A. graveolens* is also edible.

Conium maculatum★ Plate 7

HEMLOCK – MOTHER DIE 2

A tall, unpleasant-smelling, annual or biennial plant with noticeable reddish purple spots on the stems. The hairless leaves are soft and feathery, being two-pinnate to four-pinnate. Bracts taper evenly to a narrow point and are bent back toward the stem. Bracteoles are similar in shape, but smaller, and turned only slightly downward; they are present only on the outside of the sub-umbels.

A very common plant of waste areas and forest margins, it flowers from September to January.

This plant is very poisonous in all its parts, particularly the seeds, since it contains the alkaloid coniine. In ancient Greece a drink made from the seeds was used for poisoning people who had been sentenced to death, and it is said that the philosopher Socrates was executed in this way.

Daucus carota★ Plate 7

WILD CARROT 1–7

A medium to tall annual or biennial with three-pinnate, hairy leaves. The bracts of the inflorescence are large and pinnate, with narrow, pointed segments. The bracts and stalks of the umbels curl inward when the inflorescence is young, then the inflorescence becomes flatter, and finally curls inward again as the seeds ripen. The flowers are white, but there may be a few reddish or blackish purple flowers in the middle of each umbel. The spines on the fruits are a distinctive feature.

It grows on waste areas, beside roads, on cultivated land and in weedy gardens: flowers from August to May.

This is the wild form of the cultivated carrot.

Foeniculum vulgare★ Plate 7

FENNEL 1–2

A tall perennial with a strong 'liquorice-like' smell. The three-pinnate, four-pinnate or five-pinnate leaves are finely divided into thread-like segments. The plant has a greyish green colour. The flowers are yellow and there are no bracts on the umbels.

It grows in waste places, along roadsides, and on coastal cliffs: flowers from November to May.

The leaves of fennel are used as a vegetable, its roots can be grated into salad, and its seeds are used as a flavouring. In the past the seeds have been used for flavouring gin. In earlier times fennel was used by the Chinese and Hindus to treat snake-bite and scorpion-bite, and the leaves were made into wreaths by the Romans as an emblem of flattery. It was also believed that placing fennel seeds in a keyhole would keep a house free from ghosts.

Periwinkle Family – Apocynaceae

SLT OW E - R
K 5 deeply lobed
C (5) funnel-shaped or having a narrow tube opening into a saucer-shaped limb (as in periwinkle, see below)
M 5 alternate
F $\underline{(2)}$ or half-inferior

In a few members of this family, the sap contains latex and is used as a source of rubber. Seeds of *Strophanthus* spp. are used in Africa to make poisons for arrows and have more recently been used as drugs in Western medicine. Two family members are familiar as ornamental shrubs: oleander (*Nerium oleander*), with pink flowers and narrow leaves, is commonly planted; frangipani (*Plumeria rubra*), with cream or pinkish flowers and a strong sweet perfume, is grown in warmer areas.

Vinca major★ **Plate 31**

GREATER PERIWINKLE **35–40**

A short, creeping perennial with shallow rhizomes and green stems up to 2 m long, woody at the base, rooting at the tip, and then turning up. The leaves are opposite, and evergreen – usually dark green, but variegated forms are sometimes found. It forms a mat covering large areas. The flowers are solitary, with distinctive, square-cut petals, their limbs spreading out in a saucer-shaped disc. The sepals are more than 7 mm long, distinguishing this species from the lesser periwinkle, *Vinca minor*, in which the sepals are less than 5 mm long.

It grows on waste land and by roadsides, often escaping from gardens where it is grown for ground cover. It grows well under trees and shrubs, and flowers all the year round.

Cultivated in gardens as ground cover under trees and bushes, it frequently escapes and can be a troublesome weed in reserves. It was a custom in Flanders to strew the lesser periwinkle on the path of the bride and groom on their way to church, its evergreen leaves symbolising the everlasting love of the couple. In contrast to this, wreaths of this plant were worn by criminals on their way to being executed.

Daisy Family – Asteraceae

HfewST ABOfewW ESiC - RI

K 5 represented by a pappus, of hairs, awns or scales, or reduced to a rim on the top of the ovary.
C (5) see below
M 5
F (2)

This very large family has many representatives, both native and introduced, in New Zealand. The family is named after a typical genus *Aster*, but is often known as the Compositae; this name comes from the distinctive feature of all members of the family – the composite flower-heads. What appears to be a single flower is actually made up of several, perhaps hundreds, of tightly packed flowers. There are two main types of flower:

- tubular – the five petals are fused into a tube with five (sometimes three) short lobes
- ligulate – the five petals are fused and the petals are continued on one side as a strap-shaped ligule

The flowers are arranged on a swollen end of the stalk, the receptacle. There are three main ways the flowers are arranged:

- All flowers are tubular.
- All flowers are ligulate (sometimes the flowers at the edge are larger).
- Flowers at the centre are tubular (forming the disc), surrounded by a single row of ligulate flowers (forming the rays).

The flowers are surrounded by one or more rows of (usually) green bracts. In some species there may be scaly bracts on the receptacle.

In spite of its many species, this family has few members of great economic importance. Sunflower (*Helianthus annuus*) is widely grown for its edible oil. A few other members are foods:

Typical daisy flowers

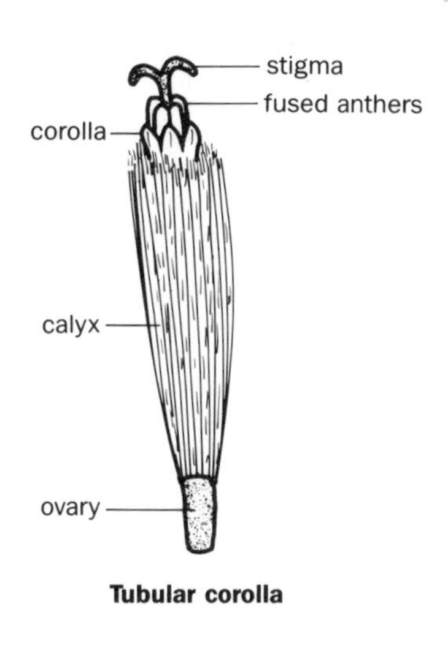

Tubular corolla

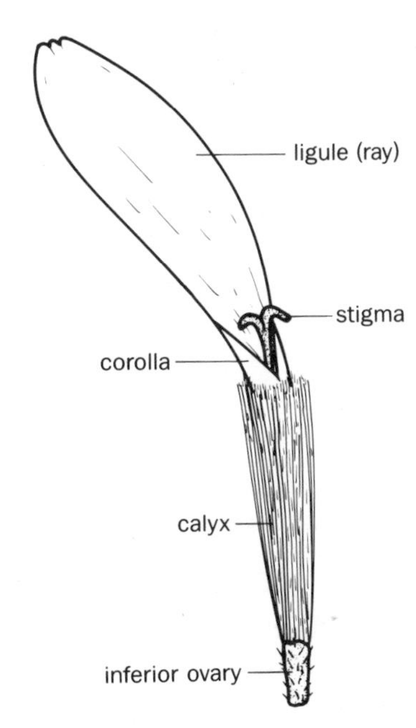

Ligulate corolla

- lettuce – leaves of *Lactuca sativa*
- Jerusalem artichoke – tubers of *Helianthus tuberosus*
- globe artichoke – the fleshy bracts of the flower-heads of *Cynara scolymus*
- endive – leaves of *Cichorium endivia*
- chicory – blanched 'hearts' of *Cichorium intybus*
- salsify, *Tragopogon porrifolius*, also known as vegetable oyster, because its roots are thought to taste like oysters

A number of species are used for flavouring, including tarragon in tarragon vinegar and tansy and chamomile in herbal teas.

The flower-heads of two species of *Chrysanthemum* are dried and powdered to provide the insecticidal dust, pyrethrum. Many members of this family are grown as ornamental plants in gardens, and of these, several have escaped from cultivation and become established in the wild. The following pages include several examples but, as well as these, we can mention Michaelmas daisy (*Aster*), golden rod (*Solidago*), marigold (*Tagetes*), cosmos, dahlia and zinnia.

For this family only, we give the diameter in millimetres of the flower-heads (including the rays, if any) instead of the diameter of the individual flowers.

The family is divided into several tribes. We use these tribal divisions here since it helps to bring close together species that are similar in appearance.

CHAMOMILE TRIBE

Members of this tribe are usually strongly scented and have alternate leaves. The flowers are usually yellow or white and there may or may not be ray flowers. The tubular flowers have short corolla lobes. The flower-head is surrounded by two to several rows of bracts. The pappus is reduced, usually to a rim on the top of the fruit, or may be absent. There is no latex.

Achillea millefolium★ **Plate 9**

YARROW **up to 10**

A short to medium, hairy, perennial plant with creeping rhizomes and erect, furrowed, woolly stems. It gives off a strong aromatic smell when crushed. The two-pinnate leaves are dark green and finely divided, the segments being partly divided pinnately, which gives the leaves a feathery appearance. The flower-heads are arranged in a flat-topped inflorescence, so that this plant can be mistaken for one of the carrot family; but each 'flower' is really a flower-head consisting of about 20 tubular disc flowers and about five ray flowers with short, off-white or pinkish ligules.

It grows on roadsides, on waste land and on cultivated land, especially in grassy places. Quite often it survives mowing as a low-growing plant on lawns: flowers from December to May.

The name 'yarrow' comes from the Greek *hiera*, meaning 'holy herb'. This plant is said to help heal wounds, and is named after the military hero of Greek legend, Achilles, who used the plant to heal the wounds of his soldiers; using the plant for this purpose is not recommended since it causes inflammation of the skin if the wound is subsequently exposed to sunlight. Yarrow leaves may be eaten as a salad vegetable when young.

Anthemis cotula★ **Plate 8**

STINKING MAYWEED **15–30**

A short to medium, annual herb with a strong and very unpleasant smell. The leaves are similar to those of yarrow, though not as feathery. The solitary flower-heads have 8–21 white ray flowers, with their ligules turning back as the flower-head matures. The very narrow, sharply pointed scales between the flowers distinguish this species from the rare corn chamomile (*Anthemis arvensis*) and from chamomile (*Chamaemelum nobile*), in which the scales are spear-shaped. In chamomile, the tips of the scales are rounded, while in corn chamomile they are pointed. To examine the scales, crush a flower-head in your hand and look for scales among the flowers.

It grows in waste places, on roadsides, in pastures and in lawns: flowers from December to March.

This weed was very unpopular in the days of hand harvesting, as its juices cause blisters on the skin. Scentless mayweed (*Triplospermum inodorum**) looks very much like stinking mayweed, but has almost no smell, has larger flower-heads (30–50 mm) and no scales on the receptacle.

Cotula australis — Plate 12

SOLDIER'S BUTTONS — 30–50

A low, annual plant, with hairy (compare with *C. coronopifolia*, below) stems and leaves. The leaves are one-pinnate or two-pinnate, deeply divided. The lower leaves are opposite, their stalks fused, forming a cup around the stem. The flower-heads are solitary on long, hairy stalks, with pale yellow flowers.

Widely occurring in waste places and other disturbed areas, it flowers all the year round.

The name *Cotula* comes from Greek, meaning 'little cup'; this refers to the bases of the lower leaves forming a little cup around the stem.

Cotula coronopifolia — Plate 12

BACHELOR'S BUTTONS – BUTTONWEED — 10

A low to short, hairless (compare with *C. australis*, above), mat-forming annual, with fleshy stems. It has a camphor-like, aromatic smell. The leaves are fleshy, alternate, their bases forming a cup around the stem, and they are not as finely divided as those of *C. australis*. The flower-heads are solitary, with very short-rayed outer flowers, making it appear that all the flowers are tubular. The flowers are a much brighter yellow than those of *C. australis*.

It grows in relatively wet areas of waste land, especially on or near the coast and around lagoons and swamps: flowers all the year round.

*Leucanthemum vulgare** — Plate 8

OXEYE DAISY – MOON DAISY — 30–60

A short to tall, slightly hairy perennial with dark green leaves, the lower leaves with long stalks, the upper leaves with none, and clasping the stem. The large, solitary, daisy-like flower-heads make this species easy to identify. The only possible confusion is with Shasta daisy (*Leucanthemum maximum*★), which has larger flower-heads (70–120 mm diameter).

It grows in grassy places on roadsides and in other waste areas, in grassland and at the margins of forests: flowers from August to May.

This is an attractive and conspicuous plant of the roadside verges. To send someone this flower means 'Be patient!'. In Europe this plant is usually known as marguerite, a name that is given to another species, *Argyranthemum frutescens*, in New Zealand.

Matricaria dioscoidea★ **Plate 13**

RAYLESS CHAMOMILE – PINEAPPLE WEED **4–10**

A short, hairless, annual plant, smelling strongly of pineapple. The much-divided feathery leaves and the rayless, conical flower-heads make this an easy plant to identify.

It grows on roadsides and in other waste and disturbed areas, and is a common garden weed, growing particularly on well-trodden paths and pavements. It flowers from October to May.

A related plant, German chamomile (*Chamomilla recutita*), is used for making chamomile tea, said to have a calming, sedative influence.

DAISY TRIBE

These are daisy-like plants, most with alternate leaves and without scent. The tubular flowers have short corolla lobes. The flower-heads have two or more rows of bracts, and there are no scales on the receptacle. The pappus usually consists of hairs, although it is reduced to scales, awns, or a rim in some species. There is no latex.

The key below includes several species that are not members of the daisy tribe, but have 'typical' daisy-like flower-heads, with white rays and (usually) a yellow disc. To simplify the wording, the key uses the term 'flowers' instead of the correct term 'flower-heads'.

IDENTIFYING COMMON DAISY-LIKE FLOWERS

1 Flowers solitary ▶ 2

Flowers in clusters ▶ 8

2 Flower-stems without leaves or scale-like bracts ▸ 3
Flower-stems with leaves or scale-like bracts ▸ 4

3 Flower-stems arising from a basal rosette of leaves ▸ ***Bellis perennis*** (p. 61)
Flower-stems branching from much-branched stems ▸ ***Erigeron karvinskianus*** (p. 64)

4 Flower-stems with narrow, scale-like bracts ▸ 5
Flower-stems with leaves or leaf-like bracts ▸ 6

5 Bracts, flowering stem and leaves covered with thin, whitish felt; rays of flowers stiff and papery, persisting on dead flowers ▸ ***Helichrysum bellidioides*** (p. 68)
Bracts, stems and leaves often green and hairless; rays of soft, petal-like texture, withering on dead flowers ▸ ***Celmisia* species** (p. 62)

6 Disc purplish ▸ ***Osteospermum fruticosum*** (p. 65)
Disc yellow ▸ 7

7 Leaves much divided and feathery; unpleasant smell ▸ ***Anthemis cotula*** (p. 58)
Leaves toothed, but not divided ▸ ***Leucanthemum vulgare*** (p. 59)

8 Leaves opposite ▸ ***Galinsoga parviflora*** (p. 66)
Leaves alternate ▸ 9

9 Robust plant, with erect stems ending in dense, many-flowered clusters ▸ ***Dolichoglottis scorzoneroides*** (p. 72)
Slender, spreading, much-branched plant, with loose few-flowered clusters ▸ ***Erigeron karvinskianus*** (p. 64)

*Bellis perennis** — Plate 8

DAISY — 20

A low, perennial plant with a basal rosette of broad, spoon-shaped leaves. The flower-heads are solitary, and their white rays are often tipped with red, especially on the outside.

It grows in waste areas, especially on lawns and in pastures: flowers from September to March.

The 'day's-eye' flower closes at night or in bad weather. The name *Bellis* is from the Latin *bellus*, meaning 'pretty'. The small size of the plant and its flowers led people to believe that the plant would make other things small; for example, at one time the roots were boiled in milk and fed to puppies to keep them small.

Celmisia allanii **Plate 8**

CELMISIA **30–40**

A low, branching shrub with rosette-like clusters of leaves at the tips of the branches. The leaves, 30–40 mm long and 10–15 mm wide, are soft and flexible, covered on both surfaces with a thin felt of white hairs. The covering on the lower surface is dense, so that the leaves appear pale grey above and intensely white below. Brown, dead leaves remain attached to older parts of the branches.

It lives in the South Island, in grassland, and herbfield at 1300 m to 1700 m: flowers in December and January.

Celmisia semicordata **Plate 8**

MATUA-TIKUMU **50–70**

A short to medium, tufted perennial, often forming large clumps. The leaves are lance-shaped, 300–400 mm long and 40–80 mm wide. The upper leaf surface has a whitish covering or 'skin' of matted hairs, which becomes detached from older leaves. The lower leaf surface is covered in a smooth, white, satiny layer of matted hairs. The midrib of the leaf is prominent, coloured orange or yellowish.

It lives in the South Island in grassland, herbfield and fellfield from 800 m to 1200 m. Sometimes found in lowland areas, it flowers in December and January.

This plant is similar in appearance to *C. armstrongii*, which has rather narrower, more rigid leaves, with a more delicate greyish 'skin' on the ribbed, yellow-green upper surface and a broad, yellowish band either side of the midrib. The edges of the leaves of *C. armstrongii* are slightly curled under.

Celmisia sessiliflora **Plate 8**

WHITE CUSHION DAISY **10–20**

A low-growing shrub forming cushions up to 1 m in diameter, with leaves closely packed and overlapping each other along the stem. Leaves are 10–20 mm long, covered on both surfaces with short, white hairs. The stalks of the flower-heads are very short, so that they are level with the tips of the branches. The stalks get longer at fruiting.

It grows in the South Island at 700 m to 1800 m, on grassland, herbfield and fellfield: flowers in December and January.

A species with similar appearance is *C. argentea*, which has smaller leaves (6–12 mm long) and flower-heads (6–12 mm diameter). The name *sessiliflora* refers to the sessile (stalkless) flowers.

The celmisias

This genus is confined to Australasia and almost all its 60 or more species are native to New Zealand; to include all of the numerous New Zealand species is beyond the scope of this book. Most of the New Zealand species grow in subalpine to high-alpine regions. Here we illustrate five typical celmisias, each from a different subsection of the genus. Identifying the species of a celmisia is a difficult matter, as so many of them look alike at first glance. The situation is further complicated by the fact that there are numerous natural hybrids.

The similarity between species eases the task of identifying a plant as 'a celmisia', but it makes it harder to distinguish one celmisia species from another. Trampers, mountaineers and others who are likely to come across the wider range of celmisias in the field should refer to the specialist books on alpine flora.

Celmisia species have very similar flower-heads, but may vary widely in their vegetative form. The flower-heads are 'daisy-like' in appearance, though larger than the common daisy described above; they are solitary on the ends of (usually) long stalks which bear numerous bracts. The disc flowers are yellow (except in *C. versicosa*, in which they are purple, but this occurs only on offshore islands). The ray flowers are white. The fruits have a pappus of rough hairs.

There are two main forms of celmisia:

- Low-growing shrubs, with leaves closely packed and overlapping each other along the stems, or in rosettes at the ends of the stems (for example, *C. allanii* and *C. sessiliflora*, described above).
- Short, perennial herbs with leaves in rosettes at the base of the stems (for example, the three other species described here).

The leaves vary in size, but are usually simple in outline and very often narrow, with a soft, white covering of short hairs, at least on their undersides and sometimes above as well. Apart from these main features, the vegetative form of the celmisias ranges from compact cushions to spreading mats, to individual, 'daisy-like' rosette plants.

Celmisia spectabilis — Plate 8

TIKUMA – PUAKAITO – PUHERETAIKO – COTTON DAISY — 30–50

A short, perennial herb with tough, leathery leaves, hairless and shiny on the upper side, covered with a thick, buff-coloured felt of hairs on the underside. This is an example of the 'daisy-like' celmisias, which grow as individual rosette plants, though also occasionally in clumps.

This species is the most widespread of the genus, and is found in grassland, tussock and fellfield from 300 m to 1700 m, in both the North and the South Island: flowers in December and January.

The spectacular flowers give this species its name *spectabilis*.

Celmisia verbascifolia Plate 8

CELMISIA **20–25**

A short to medium, perennial herb with tough, leathery leaves, which have a distinct leaf-stalk, unlike most other celmisias. The upper surface of mature leaves is hairless and shiny with prominent veins. The lower surface is covered with a dense, velvety felt of white to pale buff hairs. The edge of the leaves is curled upward, making the leaves appear as if they have a whitish/buff fringe.

It grows in the South Island (Otago and Fiordland) in rocky places in coastal areas, subalpine grassland and herbfield: flowers in December and January.

The name *verbascifolia* means 'verbascum-leaved', referring to the mullein plant, *Verbascum thapsus*, which has similar leaves.

Conyza bilbaoana★ Plate 14

CANADIAN FLEABANE **1.5–2.5**

A tall, annual or biennial plant with ridged stems and bristly hairs. The narrow leaves have no leaf stalks, the leaf blade tapering toward the stem. The leaf may have widely separated teeth or be shallowly cut into pinnate segments. As seen in the photograph, the inflorescence consists of a long main stalk, with much shorter side branches, bearing numerous small flower-heads. The appearance of the flower-heads is dominated by the green bracts, the ligules of the outer flowers being very short and hardly visible. The ligules are cream, later tinged with purple.

Found on waste land and cultivated land, it flowers from September to May.

The related species, wavy-leaved fleabane (*C. bonariensis*★), is distinguished by its twisted leaves with wavy edges.

Erigeron karvinskianus★ Plate 9

MEXICAN DAISY **15–25**

A short, perennial herb with branching, leafy stems. The lower stem leaves are often

DAISIES WITH WHITE-RAYED FLOWER-HEADS

Osteospermum fruticosum
p. 65

Erigeron karvinskianus
p. 64

Galinsoga parviflora
p. 66

Ageratina adenophora
p. 65

Achillea millefolium
p. 58

Craspedia sp.
p. 67

Helichrysum bellidioides
p. 68

Leucogenes grandiceps
p. 68

PLATE 10

DAISIES WITH YELLOW-RAYED FLOWER-HEADS

Dolichoglottis lyallii
p. 72

Hieracium praealtum
p. 81

Hieracium aurantiacum
p. 81

Crepis capillaris
p. 80

Hypochoeris glabra
p. 82

Lapsana communis
p. 82

Picris echioides
p. 83

Sonchus asper
p. 83

Sonchus oleraceus
p. 84

DAISIES WITH YELLOW-RAYED FLOWER-HEADS

Helichrysum bracteatum
p. 68

Brachyglottis bellidioides
p. 71

Brachyglottis lagopus
p. 71

Senecio jacobaea
p. 74

Brachyglottis haastii
p. 71

Senecio vulgaris
p. 75

Senecio skirrhodon
p. 75

Senecio lautus
p. 74

PLATE 12

DAISIES WITH ONLY ONE YELLOW-RAYED OR RAYLESS FLOWER-HEAD ON EACH STEM

Cotula australis
p. 59

Arctotheca calendula
p. 76

Cotula coronopifolia
p. 59

Gazania linearis
p. 77

Arctotis stoechadifolia
p. 76

Hieracium pilosella
p. 81

Leontodon taraxacoides
p. 83

Taraxacum officinale
p. 84

THISTLES, DAISIES, TEASEL

Carduus tenuiflorus
p. 77

Senecio elegans
p. 73

Cirsium vulgare
p. 78

Onopordum acanthium
p. 78

Cirsium arvense
p. 78

Cichorium intybus
p. 80

Matricaria dioscoidea
p. 60

Dipsacus sylvestris
p. 105

PLATE 14

FIREWEEDS, FLEABANE, LOGFIA

Senecio bipinnatisectus
p. 72

Senecio diaschides
p. 73

Senecio glomeratus
p. 74

Senecio minimus
p. 75

Senecio quadridentatus
p. 75

Conyza bilbaoana
p. 64

Logfia minima
p. 69

CUDWEEDS, GOOSEFOOTS, ORACHE, GLASSWORT

Gnaphalium coarctatum
p. 67

Gnaphalium involucratum
p. 67

Chenopodium pumilio
p. 100

Chenopodium album
p. 99

Chenopodium murale
p. 100

Atriplex prostrata
p. 99

Sarcocornia quinqueflora
p. 100

PLATE 16

Wireweeds, inkweed, purslane

Polygonum aviculare agg.
p. 144

Polygonum hydropiper
p. 145

Polygonum persicaria
p. 145

Polygonum strigosum
p. 146

Phytolacca octandra
p. 141

Polygonum polystachyum
p. 145

Portulaca oleracea
p. 148

three-lobed but the other leaves are entire. The flower-heads are daisy-like in appearance and size. Ray flowers may be white, pink, purplish pink or purple. The pappus consists of a row of long hairs.

It is found in waste areas, streamsides or scrub, often growing on banks: flowers from September to May.

Erigeron comes from Greek, meaning 'early/old man'; this refers to the whitish, hairy appearance of the flower-heads, seen rather early in the year. Probably this name was originally used for another plant, groundsel, but has been transferred to this one. This plant is a garden escape which has now become widely naturalised.

MARIGOLD TRIBE

These daisy-like plants mostly have alternate leaves and are usually strongly scented. The flower-heads have one to three rows of bracts, with membranous edges. There are no scales on the receptacle, no pappus and no latex. The yellow-flowered garden marigold is a member of this tribe.

Osteospermum fruticosum★ — Plate 9

DIMORPHOTHECA — **40–70**

A short, straggling or mat-forming perennial with fleshy leaves. The flower-heads are solitary. The rays are white on the upper surface, but bluish mauve below. The relatively small disc, of bluish purple flowers, often appearing almost black, is a distinctive feature of this plant.

This plant is often grown in gardens and has frequently escaped. It has become widely naturalised on banks and cliffs, especially near the coast. It flowers from August to January.

HEMP AGRIMONY TRIBE

The leaves are usually opposite, and the bracts in two to many rows. All flowers are tubular, and there are no scales on the receptacle. The pappus is of hairs. There is no latex.

Ageratina adenophora★ — Plate 9

MEXICAN DEVIL — **5–7**

A tall, perennial herb with very hairy, branching stems. Leaves are opposite, kite-shaped, with toothed edges except near the leaf base. The many flower-heads are clustered in more or less flat-topped inflorescences. What appear to be narrow ray flowers in the photograph are the styles and stigmas of the mature outer flowers.

It is found in the North Island (north and south Auckland, Coromandel) on the margins of forests and in clearings, and in waste areas. It flowers from August to December.

A common relative of this plant is mist flower (*A. riparia**). This has narrower leaves and smaller flower-heads, and is distinguished by the purple-striped hairs on the upper stem and leaf-stalks.

SUNFLOWER TRIBE

Members of this tribe usually have opposite leaves and are not scented. There are one to several rows of bracts, and scales are usually present on the receptacle. Flowers are usually yellow or white. The pappus consists of scales, awns or a ring, or is absent. There is no latex.

*Galinsoga parviflora** — Plate 9

GALINSOGA – GALLANT SOLDIERS — 3–7

A short to tall, branching annual with clusters of flower-heads at the tips of the stems. There are up to five rays on each flower-head, and these are small and white. Leaves are opposite and simple with toothed margins. Pappus scales are broad, with rounded tips.

It grows in disturbed soil, such as gardens and other cultivated areas.

The fruits formed by the disc flowers are topped with a pappus of large, broad scales up to 2.5 mm long, which aid wind dispersal. The fruits formed by the ray flowers have no pappus scales or only small ones, and will not be dispersed far. This is thought to give the plant a chance to spread widely if suitable new areas are within range, but, if not, to increase the number of plants growing close to the parent.

A similar, though less common, species is *G. quadriradiata**, which has narrow pappus scales with sharply-pointed tips.

CUDWEED TRIBE

These have alternate leaves and are usually not scented. There are several rows of bracts, which are sometimes spreading and distinctively coloured. Flowers are yellow, orange or white. The pappus is usually of hairs. There is no latex.

Craspedia species — Plate 9

PUATEA — 4 long

A medium, rosette-forming, perennial herb with sparse, rather rough hairs on the leaves and often on the flower-head stalk. As well as the basal rosette of leaves, there are smaller leaves on the flower-head stalk. The 'flower-head' is actually a tight cluster of 10 or more flower-heads, surrounded by a rosette of reduced leaves, about 4 mm long.

It is found on the coast of the South Island, especially in the fiords, growing in rock crevices and on the shore: flowers from October to March.

Six species of *Craspedia* have been described by various authors but others consider that all the forms found in New Zealand are simply variants of a single species.

Gnaphalium coarctatum★ — Plate 15

PURPLE CUDWEED — 1 long

A short to tall annual or biennial, often with a single, unbranched stem. If the stem branches, it does so only near the base. The leaves are distinctive in having a dense, purplish white woolly covering on the underside, while the upper surface of all except the youngest leaves is hairless, shiny and dark green. The inflorescence consists of a spike of clusters of tiny flower-heads, with strap-shaped leaves between them.

A common weed of waste areas, lawns, pasture and cultivated land generally, it flowers from September to March.

Gnaphalium involucratum — Plate 15

CREEPING CUDWEED — 1.5–3

A short to tall, perennial plant creeping by means of stolons. The leaves have a dense, white, woolly covering on the lower surface, while the upper surface is hairless, shiny and dark green. The inflorescence consists of a small, dense cluster of about 10–15 tiny flower-heads, surrounded by a rosette of leaves of different lengths. The length of these leaves is between one and four times the diameter of the clusters.

It grows in pasture, scrub, lawns and gardens, particularly in open, wet situations, and flowers from November to February.

Helichrysum bellidioides **Plate 9**

HELICHRYSUM **14–22**

A low to short, perennial herb with spreading, rooting stems and erect stems which each bear a solitary flower-head. The flowers are all tubular and there are no ray-flowers. Instead, the bracts around the flower-heads are elongated and spreading, looking like white rays. They are stiff and membranous in texture, persisting after the flower-heads are dead.

It occurs in a wide range of open places, at altitudes ranging from lowland to alpine: flowers from October to March.

The genus name *Helichrysum* comes from the Greek *helios* (the sun), and *chryson* (golden). Many of the genus (for example, strawflower, below) have golden, sun-like flowers and so warrant this name. This particular species has flowers more like the daisy (*Bellis*), and its species name means 'daisy-like'.

*Helichrysum bracteatum** **Plate 11**

STRAWFLOWER **25–50**

A short to medium, perennial herb with erect, leafy stems bearing clusters of one to five flower-heads. As in *H. bellidioides* above, the 'rays' are actually wide-spreading bracts, in 7–15 rows. In this species, their bright yellow, orange, gold or white colour makes the plant easy to identify.

It lives in waste areas and on forest margins: flowers from November to April.

Leucogenes grandiceps **Plate 9**

SOUTH ISLAND EDELWEISS **9–15**

A low, whitish grey or whitish buff, spreading perennial. The 'flowers' consist of 5–15 rayless flower-heads, densely clustered together and surrounded by up to 15 white woolly bracts, which are up to 10 mm long. The leaves are 5–10 mm long and 2–4 mm wide, with blunt tips.

It grows in rocky subalpine to alpine habitats in the South Island: flowers from November to March.

The North Island edelweiss (*L. leontopodium*), which occurs in both islands, is a larger plant. The woolly bracts are up to 20 mm long, and there are up to 20 of them. Its 'flowers' are up to 25 mm in diameter. The leaves are 8–20 mm long and 4–5 mm wide, with pointed tips. The basal leaves are in rosette. These species derive their common name from the fact that they are similar in appearance to the European edelweiss, *Leontopodium alpinum*.

*Logfia minima** — Plate 14

SMALL CUDWEED — **1–2**

A low to short, silvery grey, annual plant. The small, narrow leaves are spirally arranged in the stem. Flower-heads are white but appear yellow because of the sepal-like bracts. The flower-heads are tightly clustered together in groups of two to eight. On fruiting, the bracts spread out flat to make conspicuous stars, several times larger in diameter than the flower-heads.

It grows in waste places, particularly rocky and stony ground, in the South Island: flowers from November to March.

Raoulia australis — Plate 38

SCABWEED — **4–5**

A low, mat-forming, greyish, creeping perennial. Branching is dense, the branches taking root. The terminal branches are erect. The spoon-shaped leaves are closely packed, overlapping in five rows and up to 2 mm long; they have a dense, woolly covering on both surfaces, except at the leaf base. One nerve is visible at the leaf-base. Flower-heads consist of 12–20 flowers surrounded by spreading bracts about 2 mm long, with bright yellow, spreading tips.

It lives on bare ground in rocky places in lowlands and up to 1600 m: flowers in January and February.

Raoulia grandiflora — Plate 38

RAOULIA — **up to 15**

A low, cushion-forming or mat-forming, silvery grey perennial; cushions are up to about 150 mm in diameter. The stems are closely covered with small, overlapping leaves, up to 10 mm long. The flower-heads are on very short stalks, so that they lie level with the 'surface' of the cushion or mat. The flower-heads are surrounded by spreading, papery, white-tipped bracts that appear to be ray petals.

It grows on rocks and in fellfield from 1000 m to 1900 m: flowers from November to January.

This species has the name *grandiflora* because of its large 'flowers'. The related species

The raoulias

There are 20 species of *Raoulia* native to New Zealand, and a few other species of the genus are found in New Guinea. Their flower-heads are short-stalked or lacking in stalks, and are without rays. The bracts surrounding the flower-heads are often coloured (white or yellow) and have a papery texture. In many species, the flower-heads are relatively inconspicuous, but in several other species the outer bracts are long and showy, spreading out widely from the flower-heads and resembling the rays of a typical daisy-like flower-head.

The stems of raoulias branch freely and are covered with closely overlapping leaves. In this way, several of the raoulia species form firm, rounded cushions, 500 mm or more in diameter, often greyish in overall colour, and dotted with tiny flower-heads. At a distance they have the appearance of sheep, earning them their common name, 'vegetable sheep'. Other species form spreading mats on stony areas such as river-beds, and are known as scabweeds.

R. youngii is similar in appearance, but its leaves are densely covered with a white to pale buff felt, and the plant forms very soft snow-white to buff mats. Its leaves are not more than 5 mm long.

Raoulia subsericea — Plate 38

RAOULIA — **7–10**

A low, pale green perennial, forming mats 200 mm or more in diameter. The stems are covered with overlapping leaves, the lower surfaces of the leaves with a silvery to golden felt, and there is usually a tuft of felt at the leaf-tip. The flower-heads are surrounded by spreading, papery bracts with blunt white tips (not as conspicuous as those of *R. grandiflora*).

It is found in the South Island, commonly on open sites in grassland, from 400 m to 1500 m: flowers from November to January.

RAGWORT TRIBE

Members of this tribe usually have alternate leaves and are not scented. The flower-heads have one or two rows of bracts and there are no scales on the receptacle. Flowers may have rays or not, and are usually yellow. The pappus is usually of hairs. There is no latex.

Brachyglottis bellidioides **Plate 11**

BRACHYGLOTTIS **20–30**

A low to short, annual plant, with a basal rosette of leaves pressed close to the ground. The leaves are usually hairy on the margins, but there are several varieties that differ widely in the amount of hairiness of the upper and lower surfaces of the leaves.

It grows in South Island grassland, tussock and scrub from 300 m to 1800 m: flowers from October to March.

The name *Brachyglottis* means 'short-tongue', and refers to the short ligules of the ray-flowers. The name *bellidioides* means 'daisy-like', and refers to the rosette leaves and solitary (usually) flower-head on a long slender stalk.

Brachyglottis haastii **Plate 11**

BRACHYGLOTTIS **20–40**

A short to medium plant with broad, rounded leaves arising from a stock, and not pressed to the ground. When young, the leaves have a dense, white, felty covering on both surfaces; this may disappear from the upper surface on older leaves (see photograph). Similarly, the flower-head stalks are felted when young, becoming hairless later. The edges of the leaves have rounded teeth. The flower-head stalks branch slightly to produce few-flowered inflorescences.

It grows in the South Island, in lowland to subalpine grassland: flowers from December to February.

Brachyglottis lagopus **Plate 11**

BRACHYGLOTTIS **20–40**

A short to medium, annual plant, with a basal rosette of leaves, not pressed close to the ground. When young, the leaves have numerous white, silky hairs on the upper surface, but these do not persist. The characteristic netted appearance of the upper leaf surface can be seen in the photograph. The lower leaf surface is covered with a soft woolly felt.

It is found in both islands growing in grassland, tussock and scrub, especially in rocky areas, up to 1500 m: flowers from November to February.

Dolichoglottis lyallii **Plate 10**

DOLICHOGLOTTIS **40–50**

A medium herb with an unbranched stem bearing grass-like leaves with sheathing bases. The stem is topped by a loose cluster of showy flower-heads with yellow rays.

It grows in the South Island, in damp grassland, herbfield, bogs, creek beds and waterfalls from 600 m to 1800 m: flowers from December to February.

The name *Dolichoglottis* means 'long-tongue', and refers to the rays being longer than those of *Brachyglottis.*

Dolichoglottis scorzoneroides **Plate 8**

DOLICHOGLOTTIS **40–60**

A medium herb with an unbranched stem bearing fleshy, spear-shaped leaves. The margins of the leaves may have a few small teeth. The inflorescence is a loose cluster of showy flower-heads with white rays.

It grows in the South Island, in damp grassland, herbfield, bogs, creek beds and waterfalls from 900 m to 1700 m: flowers in December and January.

The above two species of *Dolichoglottis* are often found growing together; they hybridise readily, producing a plant in which the ligules are cream to pale pink.

*Senecio bipinnatisectus** **Plate 14**

AUSTRALIAN FIREWEED **2**

A tall annual (sometimes living for a few years) with stem leaves very deeply and pinnately cut into narrow segments. The segments are toothed or sometimes pinnately lobed again, as seen in the photograph. The stem is usually hairless, but may have a few hairs. The flower-heads have three to four supplementary bracts 1–3 mm long, and a single row of 8–13 narrow, inner bracts 5–7 mm long. The small flower-heads have no rays and the disc flowers are greenish yellow.

Common in the North Island in waste areas, coastal areas, pasture, forest margins and clearings, it flowers from December to June.

The senecios

This is one of the largest genera worldwide. Although most are relatively small herbs and very successful as weeds, the genus includes the large tree-daisies of Mount Kilimanjaro in Tanzania. There are 34 species of *Senecio* growing wild in New Zealand, 18 of them native. The New Zealand members of the genus are mostly herbs, though some become woody at the base and two are small shrubs. They have alternate leaves, usually simple but often pinnately lobed. There is one row of free inner bracts around each flower-head, but outside of these there may also be a few much smaller, supplementary bracts. The numbers of supplementary and inner bracts quoted in the descriptions below are typical, but plants may occasionally be found with fewer than or more than these numbers. Flowers are usually yellow, and the fruits usually have a hairy pappus.

*Senecio diaschides**

FIREWEED

Plate 14 1–2

A tall, hairless annual (sometimes living for a few years) with stem leaves very long and narrow, their margins with shallow, widely spaced teeth. The leaf bases of the lower and mid-stem leaves are often deeply cut into three parts, and clasp the stem. The flower-heads have four to six supplementary bracts 1–2 mm long, and a single row of 8–12 narrow, inner bracts 4.5–5.5 mm long. The small flower-heads have no rays and the disc flowers are yellow.

It is common in Northland, in waste areas, coastal areas, swamps and pasture: flowers from November to April.

*Senecio elegans**

PURPLE GROUNDSEL

Plate 13 2

A low to medium, hairless or slightly hairy annual, with leaves pinnately lobed, the lobes often being pinnately lobed again, as in the photograph. The flower-heads have 11–16 supplementary bracts, 2–5 mm long, and a single row of 13–14 inner bracts, 5–9 mm long. The disc is yellow, but the 12–17 ray flowers are unusual for this genus in being purple, or sometimes pink, purplish pink or white.

It grows on sand dunes and on other coastal sites, though it is sometimes found inland, growing in waste areas: flowers from August to May.

Senecio glomeratus — Plate 14

FIREWEED — **2–3**

A medium annual (sometimes living for a few years) with pinnately cut leaves, the lobes having a few coarse teeth. The leaves and stems are greyish, with a dense, white felt on the undersides of the leaves. There are 8–14 supplementary bracts 1–2.5 mm long, and 13 inner bracts, 4.5–6 mm long. There are no ray flowers and the disc flowers are yellow.

Very common in a wide range of habitats, from sea-level to 1000 m, it flowers from November to March.

*Senecio jacobaea** — Plate 11

RAGWORT — **15–25**

A medium to tall, biennial or perennial plant. The stems have a downy covering at first, but this clears, leaving the stems hairless. Young plants have a basal rosette of leaves, which usually die away before flowering. The leaves are pinnately lobed, covered with a light web of hairs, the end lobe being small and blunt. Leaf bases clasp the stem. Flower-heads are in more or less flat-topped clusters. The ligules of the 11–13 ray flowers are erect at first, then spreading. There are 3–10 supplementary bracts, 1.5–3 mm long. The 11–14 inner bracts are 3–5 mm long, with dark brown tips. The fruit bears a pappus of whitish hairs.

It grows in waste areas, pasture (especially if it is overgrazed), grassy places, and roadsides: flowers from November to April.

The plant is poisonous to stock and is listed as a noxious plant. The name of the genus is derived from the Latin, *senex*, which means 'old man'; this refers to the whitish pappus on the fruits.

Senecio lautus — Plate 11

SHORE GROUNDSEL — **10–20**

A low, spreading, hairless or slightly hairy, annual or perennial plant. The leaves are fairly fleshy and divided pinnately into lobes or teeth, about three to six on each side of the mid-stem leaves. There are 6–16 supplementary bracts, 1–3 mm long, and 10–13 inner bracts, 4–6.5 mm long. The flower-heads have 7–13 ray flowers, with relatively short ligules, 2–7 mm long.

It lives in coastal areas on cliffs, beaches and rocks: flowers throughout the year.

Senecio minimus **Plate 14**

FIREWEED **1–2**

A tall, erect annual (sometimes living for a few years) with stem leaves narrowly lance-shaped, their margins finely toothed. The leaf bases of lower and mid-stem leaves are lobed and clasp the stem. The flower-heads have three to five supplementary bracts 1–2 mm long, and eight inner bracts 4.5–6.5 mm long. The small flower-heads have no rays and the disc flowers are yellow.

Common in Northland, in waste areas, coastal areas, swamps and pasture, it flowers from November to April.

Senecio quadridentatus **Plate 14**

COTTON FIREWEED **2**

Medium annual (sometimes living for a few years) with long, narrow, parallel-sided leaves. The leaf margins are entire or with widely spaced small teeth. The leaves and stems are greyish with a dense white felt on the undersides of the leaves. There are three to five supplementary bracts 1–1.5 mm long, and 12–13 inner bracts, 6–9 mm long. There are no ray flowers and the disc flowers are yellow.

Common in a wide range of habitats, especially stony and rocky places, from sea-level to 1000 m, it flowers throughout the year.

Senecio skirrhodon★ **Plate 11**

GRAVEL GROUNDSEL **15–30**

A short to medium annual (sometimes living for a few years), completely hairless except possibly for a few hairs on the mid-veins and bases of the leaves. The leaves are fleshy, light green, narrow or spear-shaped, with slightly toothed margins. Some leaves may be cut pinnately into three to five sections on each side. Flower-heads are solitary or few in loose clusters. There are 9–17 narrow supplementary bracts, 2–3 mm long, and 18–23 narrow inner bracts, 5–7 mm long. The bright golden-yellow ray and disc flowers are distinctive.

It grows in coastal areas and waste places: flowers in December and January.

Senecio vulgaris★ **Plate 11**

GROUNDSEL **4–5**

A low to short annual, often covered with cobweb-like hairs, except on the upper surface of the leaves. The flower-heads are nearly always without rays, in loose clusters. It has 8–21 triangular or spear-shaped, black-tipped supplementary bracts, 1–1.25 mm long, and 18–21 long, narrow inner bracts, 5.5–8 mm long, also with black pointed tips. The pappus on the ripe fruits forms prominent spherical 'clocks'.

This is very common in a wide range of waste places and disturbed areas, and a frequent weed of gardens and other cultivated land. It flowers all the year round.

The seeds of this most successful weed germinate readily when exposed to light; if covered by soil, only about 15 per cent of seeds will germinate. Living seeds remain in the soil for years, without germinating, until the soil is disturbed and the seeds are exposed to light. The name 'groundsel' comes from the medieval name *grundeswyle*, which means 'earth glutton'.

ARCTOTIS TRIBE

Many members of the tribe have large, showy flower-heads with a disc and rays, the tips of the rays having three teeth. Bracts are in four to eight rows, and there are usually no scales among the flowers. The pappus consists of usually two rows of scales or bristles. Latex is present in some species.

Arctotheca calendula★ — Plate 12

CAPE WEED — **35–60**

A low to short, spreading, felted or hairy annual, with a basal rosette of pinnately cut leaves. The end segment of the leaf is larger than the other segments, as shown in the photograph. Flower-heads are solitary, with four to five rows of greenish bracts. The rays are tinged with green or purple underneath.

It grows on roadsides, beaches and waste areas, being common on verges and as a weed of lawns: flowers from October to April.

The name 'cape weed' refers to its original introduction from South Africa.

Arctotis stoechadifolia★ — Plate 12

ARCTOTIS — **70–100**

A short to medium, trailing, silvery white perennial with toothed or pinnately cut leaves. If the leaves are pinnately cut, the terminal segment is much larger than the others. The flower-heads are large and showy in a wide range of colours, including a rich yellow-orange (illustrated), a deep pinkish purple (illustrated), brick red and purple. The disc is blackish purple, and there are five rows of thin, dry, brownish bracts.

It is commonly grown in gardens, often escaping and establishing itself in waste areas, banks and roadsides. It flowers from November to March, and outside this period in sunny situations.

The name *Arctotis* means 'bear's-ear' and refers to the scales on the fruits, which are said to look like a bear's ear.

Gazania linearis★ **Plate 12**

GAZANIA **70–100**

A short to medium perennial with leaves tufted at the ends of the rhizomes. The dead leaves remain attached to the plant. The leaves are narrow and lance-shaped, dark green above and felted white below, except on the midrib; a few leaves may be pinnate toward the tip. The flower-heads are solitary, on stalks 200–300 mm long, and their yellow or orange rays with the basal black patch, spotted white in the centre, make this an easy genus to identify. The black patch may be lacking in some varieties.

A common garden plant that has successfully escaped in many areas, it grows on waste areas, cliffs, sand dunes and stream-banks. It flowers from November to February.

A related species, *G. rigens*★, has similar flowers with orange rays, and the white-spotted black patch. It can be distinguished from *G. linearis* by having trailing stems ending in tufts of leaves, shorter (50–100 mm) flower-stems, and the fact that the dead leaves fall off.

THISTLE TRIBE

Many, but not all, members of this tribe are thistle-like in appearance, with spiny leaves and sometimes with spiny stems. The flower-heads consist only of tubular flowers and the bracts usually end in a sharp spine. The pappus usually consists of several rows of feathery hairs, bristles or scales. There is no latex.

Carduus tenuiflorus★ **Plate 13**

WINGED THISTLE **7–15**

A short to tall annual or biennial covered with cobweb-like white hairs. The stems have thorny wings, and the leaves are densely white beneath. It has small (for thistles) flower-heads in dense clusters.

This grows in waste areas, on sand dunes, beside roads, in pasture and in tussock grassland: flowers in November and December.

Slender winged thistle (*C. pycnocephalus**) is similar in appearance, but has narrower wings (only 2 mm wide) between clusters of spines, compared with 2–3 mm or even up to 10 mm in *C. tenuiflorus*. Its flower-heads are shortly stalked in clusters that are not so compact as those of *C. tenuiflorus*.

*Cirsium arvense** — Plate 13

CALIFORNIAN THISTLE — **10–20**

A medium to tall perennial growing from a rhizome. It has spiny leaves, but no spines on the stems. Flower-heads are shortly stalked, in clusters, with purplish bracts. The bracts are not spiny.

It grows in waste places, on roadsides and in cultivated land, and is often found on dry terrain, such as sand-dunes, screes and stony ground. It flowers from December to February.

*Cirsium vulgare** — Plate 13

SCOTCH THISTLE — **25–50**

A medium to tall biennial with a tap root. There are long, sharp, pale spines on the leaves, winged stems and bracts. The narrow wings arise from the bases of leaves and run down the stems. The undersides of the leaves are covered with dense grey hairs.

It grows in waste areas, on roadsides and in cultivated land: flowers from November to March.

*Onopordum acanthium** — Plate 13

COTTON THISTLE — **30**

A tall biennial with a tap root, it has sharp spines on its leaves, and broadly winged stems and bracts. The bracts are covered with cottony hairs, except for the long, yellowish spines. The waisted shape of the flower-heads is a characteristic feature of this species.

It is found on roadsides, pasture, river-beds and waste areas: flowers from December to February.

LETTUCE TRIBE

Almost all members of this tribe have a basal rosette of leaves. They differ from most other members of the Daisy Family in having latex (a white fluid which exudes from the cut ends of stalks and leaves). Their flowers are all ligulate, and usually coloured yellow. Although a few of this tribe are spiny, they differ from the thistle tribe in not having spine-tipped bracts. The tribe includes the yellow dandelion-like members of the family; many of these are easily confused with each other, and this key will help to sort them out:

IDENTIFYING COMMON DANDELION-LIKE FLOWERS

1 Flower-head solitary on an unbranched stem ▸ 2
Flower-heads more than one on a branched stem ▸ 5

2 Leaves simple, with shaggy white hairs ▸ ***Hieracium pilosella*** (p. 81)
Leaves lobed or toothed ▸ 3

3 Flower-heads 12–20 mm diameter; buds nodding; leaves with triangular lobes; outer bracts not spreading ▸ ***Leontodon taraxacoides*** (p. 83)
Flower-heads 35–55 mm diameter; leaves with large, jagged teeth pointing back toward stem; outer bracts usually spreading or turned down ▸ 4

4 Bracts entirely green or reddish ▸ ***Taraxacum officinale*** **agg.** (p. 84)
Bracts with thin white margins; usually found only at 500–1900 m ▸ ***Taraxacum magellanicum*** (p. 84)

5 Leaves clasping the stem ▸ 6
Stems leafless, or leaves not clasping the stem ▸ 9

6 Leaves with large white 'pimples' ▸ ***Picris echioides*** (p. 83)
Leaves without 'pimples' ▸ 7

7 Outer row of bracts half the length of the inner bracts, or less; bracts spreading or curved back; outer flowers often with a reddish stripe beneath ▸ ***Crepis*** **species** (p. 80)
Outer bracts not distinctly shorter than inner bracts; outer flowers not reddish beneath ▸ 8

8 Leaves shiny with sharp spines, clasping the stem with rounded lobes ▸ ***Sonchus asper*** (p. 83)
Leaves dull green with softly spiny margins, clasping the stem with arrow-shaped points ▸ ***Sonchus oleraceus*** (p. 84)

9 Juice of plant is watery; no pappus on fruits ▸ ***Lapsana communis*** (p. 82)
Juice of plant is milky (latex); pappus of hairs ▸ 10

10 Unbranched, leafy stems, with lance-shaped toothed or slightly toothed leaves; pappus of brownish hairs ▶ ***Hieracium* species** (p. 81)

Stems without leaves, rosette leaves roughly hairy; solitary or very few flower-heads; pappus white or dirty white ▶ ***Hypochoeris glabra*** (p. 82)

Cichorium intybus★ **Plate 13**

CHICORY **35**

A medium to tall perennial with lower leaves toothed to pinnately lobed, upper leaves simple, margins coarsely toothed, bases clasping the stem. The flower-heads are large and sky-blue in colour. The plant does not have latex.

It grows on roadsides and on waste land, and flowers from December to March.

The roots of chicory, when dried and ground, are used as a coffee substitute for blending with coffee to modify its flavour; so-called 'Prussian coffee' was prepared in this way as long ago as 1600. The leaves of the related species *C. endivia* are eaten as a salad vegetable.

Crepis capillaris★ **Plate 10**

HAWKSBEARD **10–15**

A short to medium, hairy annual or biennial with branching stem. The leaves are shiny, some in the basal rosette, others on the stem. Basal leaves have stalks and may be toothed, but are usually deeply cut into pinnate segments, the segments being directed back toward the leaf base; they are 50–200 mm long, 10–30 mm wide. Upper leaves are narrower, less deeply cut, stalkless, and have bases that clasp the stem with arrow-shaped points. Outer flowers often have a pink or reddish stripe beneath. The outer surfaces of the bracts have short, white hairs and a few dark, glandular hairs. The inner surfaces are hairless.

This is found in a wide range of disturbed habitats, including roadsides and gardens. It flowers from September to March.

Beaked hawksbeard (*C. vesicaria*★) has hairs on both outer and inner surfaces of the bracts, rather larger flowers and leaves (100–200 mm long, 40–80 mm wide) and is often taller.

Hawkweeds

These are a distinctive sub-group of the yellow dandelion-like plants, and there are hundreds of different forms (microspecies), which are extremely difficult to identify. The main features of the group are: unbranched stems bear a few (or no) leaves, and end in a branching cluster of flower-heads or a single flower-head; flower-heads are nearly always yellow; leaves are spear-shaped, alternate, with toothed margins. The fruits bear a tuft of hairs, usually pale brown in colour. Of the three species of hawkweed illustrated in this book, king devil is the most typical.

Hieracium aurantiacum★ — Plate 10

ORANGE HAWKWEED — **15–20**

A short to medium, hairy perennial plant, with a rosette of entire leaves at the base. The stem and leaves are covered with blackish hairs. The unbranched stem bears an open cluster of 5–10 flower-heads; their orange colour distinguishes this species from other hawkweeds, except from *H.* ×*stoloniflorum*★, which is a hybrid between *H. aurantiacum* and *H. pilosella* (below), and has only one or two flowers on each stem.

It lives in grassy places and waste areas, and flowers from December to March.

Hieracium pilosella★ — Plate 12

MOUSE-EAR HAWKWEED — **20**

This low to short, perennial plant produces runners, by which it spreads rapidly. The leaves have long hairs, dense beneath, forming a greyish white felt, which gives the plant its descriptive name 'mouse-ear'. The hairs on the fruit are brittle. The solitary flower-heads, pale yellow in colour, distinguish this from other hawkweeds. The flowers often have a red stripe on the outside of the ligules.

It lives in a wide range of waste and cultivated areas, as well as in tussock grassland: flowers from October to February.

Hieracium praealtum★ — Plate 10

KING DEVIL — **10–25**

A short to medium, hairy, perennial plant producing runners, by which it spreads rapidly. The leaves of the basal rosette are bluish green in colour, with coarse hairs

2–4 mm long, usually on both surfaces. The leaf is lance-shaped, entire or slightly toothed. There are no leaves, or possibly only one or two small leaves, on the stems. The stem is branching, bearing a cluster of 10–25 flower-heads.

It is found in both islands, but is more widespread in the South Island, growing in waste places, by roadsides and in scrub: flowers from November to March.

Hypochoeris glabra★ **Plate 10**

SMOOTH CATSEAR **12–15**

A low to short annual, usually with shiny, hairless leaves. Most of the leaves are in the basal rosettes, there being only a few scale-like leaves on the stem, topped by a single flower-head. Flower-heads open only in full sunlight, and the bracts are about the same length as the rays. They enlarge at fruiting, forming a narrow, conical fruiting head. Smooth catsear and catsear (see below) are easily confused with the hawkbits (*Leontodon* species). The best way of distinguishing them is to look for scales among the flowers. Take a flower-head and rub it between your palms to separate the flowers: catsears have narrow, yellowish scales among the flowers; hawkbits have none.

It lives in grassy places, sand dunes, rocky and stony places, roadsides and waste land: flowers from November to May.

The stems are often swollen where the wasp *Aulax hypochaeridis* has laid its eggs in them.

Another common species is catsear (*H. radicata*★); this is a perennial, with roughly hairy leaves. The rays are longer than the bracts, which have a row of sharp hairs along the midrib, the hairs being darker toward the tip of the bracts. The flower-heads open in dull as well as in sunny weather.

Lapsana communis★ **Plate 10**

NIPPLEWORT **10–20**

A short to tall, hairy annual with a branching stem, each ending in a finely branched cluster of 8–12 flower-heads. The stem bears several oval leaves, which are often pinnate near their base, ending in a large oval lobe. Upper leaves are undivided but toothed, and the fruits have no hairy pappus. The plant does not have latex.

It lives beside tracks and roads, particularly under trees and bushes: flowers from December to March.

The ancient Doctrine of Signatures stated that any plant part that resembles a part of the human body may be used to cure disorders or diseases of that part; the flower buds of nipplewort are shaped similarly to the human nipple and so the plant was thought to be a cure for soreness of the nipples. The ending '-wort' is commonly used for plants that were used for cures, as in toothwort, navelwort and ragwort.

Leontodon taraxacoides — Plate 12

LESSER HAWKBIT — 12–20

A low to medium perennial with a basal rosette and an unbranched, leafless stem ending in a single flower-head; the flower-heads are nodding when in bud. Lesser hawkbit is sometimes confused with dandelion, but its flower-head is smaller and its leaves are only shallowly lobed. The outer ligules are greyish on the outside. The pappus on the fruits forms a spherical 'clock'. This plant is easily confused with the catsears (*Hypochoeris* species), and the best way of distinguishing them is to look for scales among the flowers: take a flower-head and rub it between your palms to separate the flowers – catsears have narrow, yellowish scales among the flowers, while hawkbits have none.

It grows in a wide range of habitats, including lawns, swamps, sand dunes and gravel: flowers from November to April.

The related autumn hawkbit (*L. autumnalis**) differs in having a branched stem, bearing many small leaf-like bracts, especially toward the top. Its flower-heads are erect in bud.

*Picris echioides** — Plate 10

BRISTLY OX-TONGUE — 20–25

A medium annual or biennial with bristly hairs. These hairs are prominent on the leaves, where with the white blister-like swellings they give the plant its common name. The flowers are pale yellow and surrounded by broad sepal-like spreading hairy bracts.

It grows on waste and cultivated land and flowers from January to March.

The leaves may be boiled as a potherb.

*Sonchus asper** — Plate 10

PRICKLY SOW THISTLE — 20–25

A short to medium annual or biennial, usually green or bluish green in colour. Its leaves are lance-shaped, shiny on the upper surface and with wavy spiny margins. The spines are stiff and prickly compared with those of *S. oleraceus*. The bases of the leaves closely clasp the stem, with rounded lobes.

It lives on waste land, roadsides, pasture and in gardens, and also grows in sandy places on the coast: flowers from October to March.

Sonchus oleraceus★ **Plate 10**

SOW THISTLE – PUHA **20–25**

A short to tall annual, greyish in colour compared with *S. asper.* Its leaves are pinnately lobed, almost to the midrib, the end lobe being largest. The edges of the leaves are toothed with small spines, but the spines are much softer than those of *S. asper.* The leaves have pointed lobes, loosely clasping the stem.

It lives on waste land, roadsides, grassland, dunes and beaches, in gardens and in cracks in paths and walls: flowers from November to January.

In the Middle Ages, this plant was cooked and eaten as a vegetable.

Taraxacum officinale★ **Plate 12**

DANDELION – TOHETAKE **35–50**

A low to short perennial, with a basal rosette of deeply toothed and jagged leaves. The stem is single and hollow with a single flower-head. The yellow ligules are usually coloured violet-grey outside, and the bracts curl back. This plant is well known for its conspicuous dandelion 'clocks'.

It grows on roadsides and waste areas, in pasture, grassland and gardens, and on lawns: flowers from September to May.

The native dandelion (*T. magellanicum*), is smaller. The bracts around the flower-heads have thin, white margins, and do not usually curl back. It lives in mountain areas from 500 m to 1900 m.

The name 'dandelion' comes from the French *dent-de-lion*, referring to the jagged 'lion's teeth' of the leaves. The leaves can be eaten, preferably before the flowers appear, when they taste too bitter; they have a high content of nutrients, especially of vitamin A and iron. The flowers are used for dandelion wine. The roots may be eaten raw or cooked, and can be roasted and ground to make a coffee substitute. (Some people declare that dandelion coffee is indistinguishable from real coffee, and it has the advantage that it lacks the possibly harmful substance, caffeine.)

This plant, like several others of the same family, sets its seeds without fertilisation: it is apomictic, which makes cross-fertilisation impossible. One result of this is that there are hundreds of slightly different forms of dandelion, each breeding true.

Borage Family – Boraginaceae

HfewS A Si – R
K 5 may be joined at the base
C (5) see below
M 5 alternate
F (2)

Plants of this family almost all have rather bristly hairs. The inflorescence is typically coiled (see the photograph of water forget-me-not). The petals are usually fused into a tube, which may end in five shallow lobes or in a flattened disc of five rounded lobes. A feature of this family is that there is often a coloured scale on each petal at the 'throat' of the flower. Another feature of many members of this family is that the flowers change colour as they mature, usually from pink to blue.

Several of the Borage Family are grown as ornamental plants, including heliotrope (*Heliotropium*), forget-me-not (including Chatham Island forget-me-not), viper's bugloss, alkanet (*Anchusa*), lungwort (*Pulmonaria*), borage and comfrey.

*Borago officinalis** — Plate 18

BORAGE — **20–25**

A short to medium, annual or biennial, bristly-hairy plant with wavy-edged leaves. The flowers are loosely clustered and are distinguished by the purplish black stamens projecting from the centre of each flower.

Found on roadsides and waste land, usually as a garden escape, it flowers from September to May.

Borage can be used as a salad, in soups, or cooked and eaten like spinach. The flowers may be floated in cocktails to garnish them. The name *Borago* comes from the Latin *burra*, a shaggy garment – this refers to the plant's dense covering of bristly hairs.

*Echium vulgare** — Plate 18

VIPER'S BUGLOSS — **15–20**

A short to tall, annual or biennial plant with flowers in a dense branched spike. Stems and leaves are densely covered with stiff hairs. The flowers turn from pink to bright blue as they mature. Stamens have long pink filaments and project far outside the tube of the flower.

This is a very common weed of roadsides and waste land. It flowers from November to January.

The fruits consist of four nutlets which look rather like a snake's head. For this reason it was formerly used as a cure for snake-bite. The related plant, Paterson's curse (*E. plantagineum*★), is common in the North Island, rare in the South Island. Its flowers are larger, and only two of the stamens project from the tube.

Forget-me-nots

There are 40 species of forget-me-nots living in New Zealand, 34 of them native. All the native species are uncommon to rare, many being restricted to mountain areas or to one or two offshore islands; most have white or yellow flowers. The two forget-me-nots described below are representative of the introduced species.

Myosotis scorpioides★ — Plate 18

WATER FORGET-ME-NOT — **5–10**

A short to medium perennial covered with short hairs pressed closely to the stem and sepals. The teeth of the calyx are one-third its length when the flower bud opens, and the ends of the petals are notched.

Common in wet places, it flowers from November to March.

The name *scorpioides* refers to the way the inflorescence curls round like the tail of a scorpion. For this reason the plant has also been called 'scorpion grass'. The other water forget-me-not, *M. laxa*★, is even more common, flowering earlier from September; it has a smaller flower, 2–5 mm in diameter and the teeth of the calyx are half its length when the buds open.

Myosotis sylvatica★ — Plate 18

GARDEN FORGET-ME-NOT — **6–10**

A short to medium biennial or perennial covered with short, spreading hairs. The hairs on the sepals are hooked and the ends of the petals are rounded.

It is found as a garden escape, growing in waste places, roadsides and scrub, and as a garden weed: flowers from March to November.

In the language of flowers, the forget-me-not means exactly what its name implies. It is said that, after the Battle of Waterloo, forget-me-nots grew up over the whole battlefield, in remembrance of the men who died there.

*Symphytum ×uplandicum** **Plate 18**

RUSSIAN COMFREY **12**

A tall, roughly hairy, clump-forming perennial. The leaves are broadly spear-shaped, the blade of the leaf continuing into two narrow wings which run down the stem.
Found on roadsides and in waste areas, it flowers from November to April.

Russian comfrey is a hybrid between common comfrey (*S. officinale**) and rough comfrey (*S. asperum**), both of which occur in New Zealand, but are uncommon. The species name is from Uppland, in Sweden, which is presumably where the hybrid originated. In the Middle Ages, common comfrey was used as a remedy for broken bones; its name *Symphytum* comes from the Greek *symphysis* (growing together of bones) and *phyton* (a plant). The leaves were also eaten, after being dipped in batter and fried.

Mustard Family — Brassicaceae

HfewS A Si - R
K 2+2
C 4 upper part of petals (the limb) usually broad with the lower part (the claw) very narrow; limbs spread in typical 'cross' formation
M 2+4 outer two short; inner four long
F (2) see below

This family is named after the genus *Brassica*, which includes many cabbage-like plants. The family is also known as the Cruciferae ('cross-bearing') because of the typical flowers with four petals spreading to form a cross.

The basic structure of the ovary is the same for all members of the family, and the fruit is usually a dry capsule that splits open to release the seeds. The ripe fruit has one of two distinct shapes that are useful in identification:

- silicule – less than three times as long as broad
- siliqua – more than three times as long as broad

One form of siliqua is the lomentum, which, instead of splitting lengthwise, breaks transversely into several segments, each containing a single seed (see sea radish).

The family includes many members cultivated as food crops, chief among these being the species *Brassica oleracea*. Of this there are many varieties, which are of widely differing appearance even though they are all the same species; these are cabbage, cauliflower, brussels sprouts, broccoli, calabrese, kale, kohl rabi and red cabbage. The so-called 'flowering cabbage', a variety with pink and yellow variegated leaves, is widely planted in Japan for ornament. Other food plants in this family are turnip (*B. rapa*), mustard (*Sinapis alba*), watercress, garden cress (*Lepidium*) and horseradish (*Armoracia*). Rape seed (*Brassica napus*) is widely grown as a field crop for its rape-seed oil and as fodder. The family also includes a number of common flower-garden plants, such as wallflower (*Cheiranthus*), stock (*Matthiola*), candytuft (*Iberis*), aubretia and alyssum. The Ancient Britons cultivated woad, *Isatis tinctoria*, because of its blue dye.

Brassica rapa★ — Plate 34

WILD TURNIP — 20

A tall annual or biennial with bright green, pinnately lobed lower leaves. The upper leaves are narrowly triangular, their bases clasping the stem, and the flowers are in a flat-topped spike. The fruit is a siliqua.

It grows in pastures, gardens, roadsides and waste areas: flowers from September to February.

Cakile maritima★ — Plate 34

SEA ROCKET — 8

A short to medium, hairless annual. Stems and the pinnate leaves are succulent. The fruit is a silicule.

It lives on sandy beaches in North Island: flowers from October to January.

The two horns at the lower end of the fruit (see photograph, just below centre) distinguish it from the other species of sea rocket, *C. edentula*★, which lives mainly on stony beaches.

*Capsella bursa-pastoris** Plate 34

SHEPHERD'S PURSE 2–3

A low to medium annual or biennial with toothed, serrated or entire leaves in a basal rosette. The stem leaves are spear-shaped and clasp the stem. The flowers are clustered at the top of the inflorescence, with the characteristic heart-shaped silicules below.

It lives on waste and cultivated land, and flowers from September to January, although it can be found in flower all the year round.

This plant is also known as 'pickpocket' and 'pickpurse' because it robs the farmer of the nutrients of the soil.

Cardamine debilis Plate 34

BITTERCRESS 3–4

A short to medium perennial, with pinnate leaves. The stem is shiny green and hairless. The leaves are pinnate and slightly hairy. The leaflets are rounded, the end leaflet being a little larger than the others, of which there are one to two pairs. The leaflets have distinctly pointed lobes (note that the leaf at bottom left of the photograph does *not* belong to this plant), and are not quite symmetrical at their base. The siliquas spread out sideways from the stem.

It lives mainly in forests, from lowland to subalpine areas: flowers from October to January.

*Cardamine hirsuta** Plate 34

HAIRY BITTERCRESS 3–4

A low to short, hairy annual with a basal rosette of pinnate leaves. There are two to three pairs of lobed leaflets, the leaflets being slightly larger toward the end of the leaf. Stem leaves are similar, though with narrower leaflets. The siliquas are long and erect, over-topping the flowers.

It grows in waste areas, and flowers from August to December.

*Coronopus didymus** Plate 34

TWIN CRESS 1

A low, creeping annual or biennial with a pungent smell. The pinnately cut leaves with narrow, sharply pointed segments are characteristic, as is the way the silicules are deeply cut into two lobes. The silicule stalks are shorter than the silicules.

It grows on waste areas, roadsides and cultivated land, and on shingly sites near the coast: flowers from November to January.

A related species is wart cress (*C. squamatus**) which has more coarsely divided leaves; the silicules are not two-lobed, and they have rough, warty walls. The silicule stalks are longer than the silicules.

Diplotaxis muralis★ — Plate 35

WALL ROCKET — **8–15**

A short to medium annual or biennial with a basal rosette or pinnately lobed leaves, and usually no stem leaves. The siliquas are longer than their stalks and held at an angle to the main stem.

Found on waste land, on roadsides, in gardens and on sand dunes, this plant flowers from October to May.

Another name for wall rocket is 'stinkweed', because it gives off a smell like hydrogen sulphide gas ('rotten eggs') when bruised.

Hesperis matronalis★ — Plate 34

DAME'S VIOLET — **15–20**

A medium to tall perennial with simple, toothed leaves. Flowers are white or violet and fragrant. The siliquas are curved and up to 100 mm long.

It lives on waste land and is also found in old gardens: flowers from November to April.

The fragrance is stronger at night and the flowers are pollinated by night-flying moths. This plant was brought to England from Europe in the 16th century. Its name in French was *violette de Damas*, meaning 'violet from Damascus', which was misunderstood as *violette des dames* and so it was called, in English, 'dame's violet'.

Lunaria annua★ — Plate 35

HONESTY — **15–20**

A tall, biennial plant with branching, hairy stems. The leaves are heart-shaped, dark green and with coarsely toothed edges. The large elliptical silicules, papery and white-translucent when mature, confirm the identity of this plant.

It lives in waste areas around gardens: flowers in October and November.

The name *Lunaria* comes from the moon-like appearance of the ripe fruits.

Notothlaspi australe — Plate 35

NOTOTHLASPI — 6–10

A low, spreading, fleshy, hairless or slightly hairy perennial. It branches freely, forming a clump covered with many rosettes of leaves, which may be reddish in colour. The abundant flowers are showy and strongly scented.

It is found in the South Island, in rocky places and screes, from 750 m to 1800 m: flowers in December.

Raphanus raphanistrum subsp. *maritimus** — Plate 35

SEA RADISH — 25–30

A medium perennial, with pinnately lobed lower leaves, the terminal lobe being narrower than the four to eight pairs of closely set lateral leaflets. The upper leaves are simple and spear-shaped, sometimes with a pair of lobes at the base. The plant is rough to the touch, with stiff hairs. The petals may be yellow or white, the branching purple veins being a distinctive feature. The siliqua (a lomentum, p. 88) narrows to a point at the far end and is strongly constricted into two to six segments, each containing one rounded seed.

It grows on roadsides and on waste land near the coast: flowers from December to March.

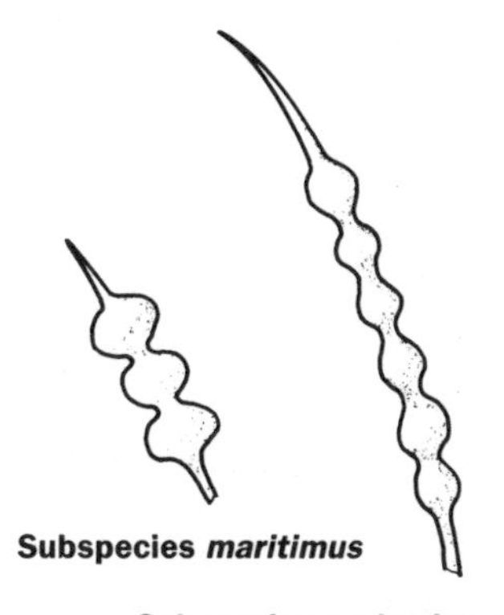

Siliquas of sea radish (actual size)

The other subspecies is wild radish *R. r.* subsp. *raphanistrum**, which grows inland and flowers from October to April; its pinnate lower leaves have one to four pairs of lateral leaflets, more widely spaced, and the terminal leaflet is the widest. The flowers may be yellow, lilac or white, with purple veins. The siliqua has more (five to eight) segments, longer than wide, with a longer, narrow region between them, and tapers more gradually to a point.

*Rorippa nasturtium-aquaticum** — Plate 35

WATERCRESS — 2

A short to medium, creeping perennial with pinnate leaves. The stem is hollow and juicy, up to 1 m long, often with small roots growing from the axils of the leaves. The small flowers are in a tight cluster at the tips of the stems, with siliquas spreading sideways below. Siliquas are 2–2.5 mm wide, their stalks are 12–15 mm long, and they contain two rows of seeds.

It grows in streams, drains and ditches, and is most common in the North Island: flowers from November to February.

This plant is rich in vitamin C. One-rowed watercress, *R. microphylla**, has larger flowers, siliquas about 1.5 mm wide, their stalks 12–20 mm long, and they contain only one row of seeds. Old leaves turn purplish brown. This is commoner in the South Island, where it flowers from November to April. A hybrid between the two species also occurs, and may be recognised by its deformed siliquas with few or no seeds.

*Rorippa sylvestris** — Plate 35

CREEPING YELLOWCRESS — 5

A short to tall perennial. Leaves are pinnate, with narrow, slightly toothed lobes; the upper leaves may be entire and slightly toothed. The petals are distinctly longer than the sepals. The siliqua is about as long as its stalk when the fruit is mature.

It grows in damp places, and flowers from December to February.

*Sisymbrium officinale** — Plate 35

HEDGE MUSTARD — 3

A medium to tall, erect, hairy annual with a basal rosette of pinnately lobed leaves. The stem is stiff and branches freely. Stem leaves may be pinnate, with spreading, backwardly pointing lobes, or arrow-shaped, with a pair of pointed lobes at the base. The cluster of flowers at the apex of each branch is inconspicuous. The stem of the mature plants bears numerous siliquas on short stalks; they are 10–15 mm long, erect and pressed very close to the stem – a distinctive feature of this plant. The siliquas have three ribs on each side.

This is found on roadsides and in waste places. It flowers from October to January.

Oriental mustard, *Sisymbrium orientale*★, is another common member of this genus. It has flowers 7 mm in diameter, and the upper leaves are not pinnate. Its siliquas are very long (40–100 mm) and spread out widely from the stem.

Bellflower Family – Campanulaceae

H A Si – R
K (5) united with the ovary
C (5) usually blue or violet; forming a bell-shaped corolla
M 5
F (2–5) occasionally superior

Both native and introduced species of this family grow in New Zealand. Of the introduced species, many were originally introduced as flower-garden plants.

Campanula rapunculoides★ — Plate 27

CREEPING BELLFOWER — **20–35 long**

A medium to tall, hairy, perennial plant with stem leaves varying from heart-shaped, to oval or narrowly oval, long-stalked below and with short stalks or unstalked above, often bent slightly downward. In the inflorescence, the nodding flowers all face to the same side. The edges of the petals have fine hairs.

A garden escape, found on roadsides and waste land near to habitation, it flowers from September to February.

Wahlenbergia albomarginata — Plate 27

HAREBELL — **15–30**

A low to short, perennial herb with a branching rhizome, each ending in a rosette of spoon-shaped or spear-shaped leaves. The leaves are up to 20 mm long, often purplish, and have thickened, white margins. A thin flower stalk up to 250 mm long grows from each rosette, ending in a solitary flower. The calyx lobes are much shorter than the corolla tube.

It lives in subalpine and alpine herbfields and grassland, and sometimes on screes, up to 1400 m: flowers from November to February.

The size of the leaves and the length of the flower-stalk is very much reduced in dry habitats.

Wahlenbergia gracilis — Plate 27

RIMU-ROA — **20**

A short or medium, annual to perennial herb with a tap root, giving rise to slender, much-branched stems. Basal leaves are usually in a rosette, spoon-shaped to narrowly oval. The upper leaves are narrower, as in the photograph, and may be opposite. The upper stem and calyx are hairless. The flower stalk is long, unbranched or slightly branched. Flowers have a short tube compared with *W. albomarginata*.

It lives in lowland and subalpine grassland: flowers from September to April.

Honeysuckle Family – Caprifoliaceae

STfewL O Si +– RI
K 4–5 or (4–5)
C (4–5) sometimes two-lipped
M (4 or 5)
F (3–5)

Although the members of this family are woody, we include the family because the honeysuckle is so common in New Zealand as a climber, and it has thin, green stems when young.

As well as honeysuckle, other ornamental plants in this family include *Viburnum*, snowberry (*Symphoricarpos*), *Weigela* and *Diervilla*.

*Lonicera japonica**

Plate 19

JAPANESE HONEYSUCKLE

20–45 long

An evergreen or semi-evergreen climber with purplish stems; leaves produced in spring are more deeply lobed than as shown in the photograph. The flower is two-lipped, the upper lip being four-toothed and the lower lip entire. The lower lip curls strongly backward. The flowers are white when they first open, but later become yellow.

It escapes on to roadsides near habitation, and flowers from September to May.

Pink Family – Caryophyllaceae

H O Si +– R
K 4 or 5 may be united at the base
C 4 or 5 sometimes absent
M 5+5 sometimes fewer
F $\underline{(2–5)}$ as many styles as carpels: carpels are opposite the petals

The stems of this family are often swollen at the bases of leaves and branches. Branching is often dichotomous, that is, a stem branches into two equal branches, which in turn branch into two more equal branches, and so on.

The family includes a few ornamental garden plants, such as pinks, carnations, and sweet William (which are all species of *Dianthus*), as well as *Lychnis*, *Silene* and *Gypsophila*.

*Cerastium fontanum**

Plate 20

MOUSE-EAR CHICKWEED

6–7

A low to short perennial, often creeping, though sometimes erect. As well as branches ending in an inflorescence, there are leafy, non-flowering branches. The petals are notched at the tip, slightly or deeply, but not for more than one-third of their length. The petals equal the sepals in length. The sepals have non-glandular hairs (that is, ending in a fine point, not a rounded 'knob'). The tips of the hairs do not project beyond the tips of the sepals. There are four to five styles, less than 2 mm long.

It lives in grassy places, on roadsides and in waste areas, scrub, sand-dunes and the banks of streams, and flowers from November to April.

There are several other common species in this genus. Snow-in-summer (*C. tomentosum*★) is covered with silvery, felted hairs. Its petals are about twice as long as the sepals, as they are also in field chickweed (*C. arvense*★). The latter occurs only in the South Island, lacks the silvery felting, and has short glandular hairs on its sepals. All the other species have petals about as long as the sepals. Annual mouse-ear chickweed (*C. glomeratum*★) has both non-glandular and glandular hairs on its sepals. The non-glandular hairs are long, and some project beyond the tips of the sepals. Little mouse-ear chickweed (*C. semidecandrum*★) lives mainly in dry places and is commoner in the South Island; its sepals have glandular and non-glandular hairs, but these do not project beyond the tips of the sepals.

Dianthus armeria★ — Plate 20

DEPTFORD PINK — **8–15**

A medium, hairless or softly hairy annual or biennial, dark green in colour (in contrast to the greyish green of many other pinks), with intensely pinkish red flowers. The flowers are in dense clusters of 2–10, closely surrounded by long, leaf-like bracts.

It lives in grassy places, on roadsides and on sand dunes: flowers from November to March.

Lychnis coronaria★ — Plate 20

ROSE CAMPION — **20–30**

A short to tall perennial with a woolly covering of silvery hairs. The flowers are in open clusters of 7–15.

It lives on roadsides and grassy areas: flowers from December to February.

The species name *coronaria* refers to the use of this plant in making garlands and crowns of flowers.

Polycarpon tetraphyllum★ — Plate 21

FOUR-LEAVED ALLSEED — **1**

A low annual with leaves appearing to be in whorls of four. The stems and leaves may be green, but are often reddish or purplish. Dry, scaly stipules are prominent. Leaves are broadly oval, sometimes almost circular in shape. Stems end in clusters of very small white flowers; the petals are shorter than the sepals.

It grows in dry habitats including roadsides, and gravelly or sandy areas: flowers from October to March.

DOCKS, PLANTAINS, NETTLE, RASPWEED

Rumex acetosella
p. 146

Rumex crispus
p. 146

Rumex conglomeratus
p. 146

Rumex sagittatus
p. 147

Plantago major
p. 142

Urtica urens
p. 172

Haloragis erecta subsp. *erecta*
p. 125

Plantago lanceolata
p. 142

PLATE 18

BORAGE FAMILY

Symphytum ×uplandicum
p. 87

Myosotis scorpioides
p. 86

Borago officinalis
p. 85

Myosotis sylvatica
p. 86

Echium vulgare
p. 85

Lonicera japonica
p. 95

Passiflora edulis
p. 140

Clematis paniculata
p. 150

Passiflora mollissima
p. 140

Calystegia sepium
p. 103

Clematis vitalba
p. 150

Calystegia soldanella
p. 103

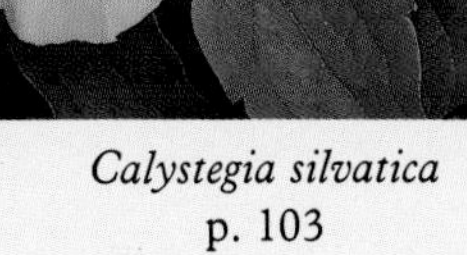

Calystegia silvatica
p. 103

PLATE 20

PINK FAMILY

Stellaria graminea
p. 98

Stellaria media
p. 98

Spergula arvensis
p. 98

Cerastium fontanum
p. 95

Dianthus armeria
p. 96

Silene gallica
var. *quinquevulnera* p. 97

Silene gallica
p. 97

Lychnis coronaria
p. 96

Pink family

Polycarpon tetraphyllum
p. 96

Sagina procumbens
p. 97

Calandrinias, lobelias, pimpernels

Calandrinia compressa
p. 147

Lobelia erinus
p. 131

Lobelia anceps
p. 130

Calandrinia menziesii
p. 147

Anagallis arvensis
p. 149

Anagallis arvensis subsp. *arvensis*
var. *coerulea* p. 149

PLATE 22

ST JOHN'S WORTS, STONECROPS

Hypericum perforatum
p. 102

Hypericum androsaemum
p. 101

Cotyledon orbiculata
p. 104

Sedum album
p. 105

Sedum acre
p. 104

SPURGES, SUNDEWS, HEATHS

Euphorbia helioscopia
p. 109

Euphorbia peplus
p. 109

Drosera arcturi
p. 106

Drosera binata
p. 106

Erica lusitanica
p. 107

Drosera stenopetala
p. 107

Gaultheria crassa
p. 108

PLATE 24

PEAS, TREFOILS, MEDICKS

Lathyrus latifolius
p. 111

Lotus suaveolens
p. 112

Lotus pedunculatus
p. 111

Lathyrus tingitanus
p. 111

Parochetus communis
p. 114

Medicago arabica
p. 113

Medicago lupulina
p. 113

Medicago sativa
p. 113

Sagina procumbens★ — Plate 21

PROCUMBENT PEARLWORT — **1**

A low, procumbent (lying on the ground), perennial plant. Narrow leaves end in a long bristle. Petals are usually four, occasionally five (both forms can be seen in the photograph). Petals are much shorter than the sepals, and may be absent. The sepals spread out when the fruit ripens, and at the same time, the thin flower-stalks curl back on themselves.

Pearlwort (*S. apetala*★) is a more upright plant, usually without petals, and often with purplish sepal margins. *S. subulata*★ usually has five petals and sepals, the petals and sepals being equal in length. A related plant was used as fodder for sheep, which is why the genus is named *Sagina*, the Latin for 'fodder'.

Silene gallica★ — Plate 20

CATCHFLY — **10–12**

A short to medium annual, covered with long sticky hairs. The flowers all point in the same direction (the inflorescences are one-sided). Petals are notched, white or pink, the flowers have 10 stamens and three styles, and the flower-stalks are 2–10 mm long.

It is widespread on roadsides, cultivated land, waste land, coastal scrub, and in gardens: flowers from October to February.

A plant of similar appearance, found in the Wellington area, is *S. disticha*★, which has the one-sided inflorescences in pairs, with a single flower growing from the axil between them. Two common species of this genus have unisexual flowers, with either 10 stamens or five styles; these are red campion (*S. dioica*★), which may also have pink flowers, and white campion (*S. latifolia*★). They have flowers pointing in all directions.

Silene gallica var. *quinquevulnera*★ — Plate 20

CATCHFLY — **10–12**

A commonly found variety of catchfly in which each petal has a dark red spot.

Spergula arvensis★ **Plate 20**

SPURREY **4–8**

A short to medium annual, covered with sticky hairs. The leaves are very narrow, with blunt tips and apparently arranged in whorls of 10 or more. Petals are about equal in length to the sepals, not notched; stamens 5–10, styles five.

It lives mainly in dry places such as sand dunes, and on bare soil in gardens and on roadsides, and flowers from September to May.

Stellaria graminea★ **Plate 20**

STITCHWORT **5–12**

A short perennial with hairless, square stems. The opposite leaves have fused bases. Petals are split to the base, and are equal to or rather longer than the sepals. The sepals have three distinct veins. There are up to 10 stamens, and three styles.

It lives in damp places such as swamps, beside streams and in wet grassy areas: flowers from December to April.

Another square-stemmed stitchwort is bog stitchwort (*S. alsine*★) with smaller flowers (4–6 mm) and petals shorter than the sepals.

Stellaria media★ **Plate 20**

COMMON CHICKWEED **8–10**

A low to short annual. The stems are round, with a distinct line of hairs running along them. The opposite leaves are not fused at the base. Petals are split to the base, and about equal to the sepals, which do not have distinct veins. There are three styles.

It lives in bare soil and cultivated areas, on roadsides and on the coast: flowers from September to February.

Goosefoot Family – Chenopodiaceae

H A Si – R
P (2), (3) **or** (5) all segments sepal-like
M as many as perianth segments; opposite to perianth segments
F (2–3)

Many of the plants in this family inhabit coastal areas and show adaptations to survival under dry conditions, often in areas where soil water is excessively salty. Their leaves are often fleshy, and are sometimes swollen and rounded. In other genera the leaves are reduced to scales. Many of the Goosefoot Family are covered with short hairs that give the plants a characteristic mealy appearance. The flowers are small and inconspicuous.

The sea beet (*Beta vulgaris*) has been bred to produce a number of varieties of economic importance. These include beetroot, sugar-beet, mangold and spinach beet. Sugar-beet is the most important source of sugar in temperate regions of the world. Spinach (*Spinachia oleracea*) also belongs to this family.

Atriplex prostrata★ — Plate 15

ORACHE — **0.5–0.8 long**

A low, spreading annual, branching profusely, with ribbed stems up to 500 mm long. Leaves and stems have a slightly mealy appearance; the middle and lower leaves vary in shape but are mostly arrow-shaped, as seen in the photograph, with the leaf base more or less square-cut. Leaf stalks are up to 30 mm long.

Often found in coastal areas, growing in the hollows in sand dunes, on mud flats, and on the banks of tidal streams, it flowers from December to February.

Another common orache is *A. patula*★, which has narrower, lance-shaped middle and lower leaves, with leaf stalks up to 15 mm long, and the leaf base tapering more gradually to the stalk.

Chenopodium album★ — Plate 15

FATHEN — **2**

A medium to tall annual with lance-shaped or rhomboid leaves, sometimes with serrated edges. The plant is a dark green colour, but the mealy covering gives it an overall bluish green colour, especially on the younger parts. The inflorescence is mealy white, and the leaf shape is very variable.

It is common in inhabited and cultivated areas: flowers from December to May.

One of the features that make this plant a successful weed is that its seeds can remain buried for a long time in the soil, yet are able to germinate when conditions are right. Seeds of fathen have been found buried at an archaeological site known to be 1700 years old; they were still able to germinate. This plant was one of the earliest food crops, and in Neolithic times and in the Bronze and Iron Ages, its seeds were ground to make flour for bread and gruel. The name *Chenopodium* means 'goose-foot' and refers to the shape of the leaf.

Chenopodium murale★ — Plate 15

NETTLE-LEAVED FATHEN — **2**

A medium to tall, mealy, erect, perennial plant, sometimes with a reddish tinge. The leaves are either green on both surfaces or purple on both surfaces. The inflorescence is green or reddish purple.

It grows in waste areas, on cultivated land and on roadsides: flowers from December to May.

A similar plant is jagged fathen (*C. erosum*★), a spreading, annual plant with leaves that are often dark purple below (but rarely above). Its inflorescences are often dark red.

Chenopodium pumilio★ — Plate 15

CLAMMY GOOSEFOOT — **1–2**

A short to medium, spreading annual, with glandular scales and aromatic smell and no mealy covering. The leaves have fine hairs and are sometimes coloured purple on the under-surface. The inflorescence is green, becoming white when the fruits are forming.

It grows in waste and cultivated areas, and occasionally on sandy coastal areas: flowers from December to March.

Sarcocornia quinqueflora — Plate 15

GLASSWORT — **less than 1**

A low to medium, shrubby perennial with succulent, opposite, scale-like leaves closely pressed to the stem and fused to surround it completely; this gives the branches a characteristic 'jointed' appearance. Leaves are translucent and green or reddish, but they may be encrusted with mud or salt. The minute flowers are in short-branched, 'jointed' spikes.

It is common on salt marshes and shingly beaches, where it is completely covered at high spring tides: flowers from November to March.

St John's Wort Family – Clusiaceae

HTS O E – R
K 2–10 or (2–10) small bracts close below calyx
C 3–12
M many or (many) if united, then often in bundles
F (3 or 5)

Until recently, this family was known as the Hypericaceae. It is also known as the Guttiferae, or 'drop-bearers', owing to drops of 'blood' (red sap) that appear when the stems are broken. Many species have clear spots on the leaves – these are oil glands. The veins of the leaves may also be translucent owing to oil channels in them.

Some of the family are tropical timber trees, also valuable for their resin. Mangosteen from *Garcinia mangostana* is an important fruit in tropical Asia. Several varieties of *Hypericum* are cultivated as ornamental shrubs.

*Hypericum androsaemum** — Plate 22

TUTSAN — **15–25**

A medium to tall shrub with two wings down opposite sides of the stem. The stems are often reddish in colour, the petals are about the same length as or a little longer than the sepals, and the stamens are many, in five bundles. There are three styles. The fruit is rounded and fleshy, reddish, becoming black when fully ripe.

It lives in scrub, in open forest areas, and in waste areas: flowers from November to February.

Its name comes from the French *toute-sain*, meaning 'all heal'; it had the reputation of healing a variety of conditions and its leaves were often placed on open wounds.

*Hypericum perforatum** — Plate 22

PERFORATE ST JOHN'S WORT — **20**

A medium, perennial herb, with two narrow ridges down opposite sides of the stem. Leaves have translucent dots, looking like pin-holes. The petals have tiny black dots (glands) on them, particularly on their margins; these may be present on the sepals too and, if so, are scattered, not just on the margins. The leaves have no black glands.

It is very common on roadsides, waste areas, and grassy places: flowers from December to May.

The common name of this plant comes from the fact that, in Europe, it flowers on or around St John's Day, 24 June. When crushed, the flowers exude a red sap and this association with blood made the plant an important one in folklore. In many countries it was the custom to hang a sprig of St John's wort over the door of the house on St John's Day, to ward off evil spirits. For the same reason it was hung above pictures; this custom gives rise to its scientific name, from the Greek *hyper* (above) and *eikon* (a picture). It was also called the 'fairy herb', because it was credited with the power to heal so many different diseases. Superstition aside, this plant is one of the most toxic pasture weeds in New Zealand.

Bindweed Family — Convolvulaceae

HS A Si – R
K 5 sometimes united
C (5)
M 5 alternate
F (2)

Most of the members of this family are climbing plants. When climbing, their stems twist anti-clockwise (that is, with the sun, as seen from above).

The most important genus in this family is *Ipomoea*. The sweet potato is the tuber of *I. batatas*. Other species of *Ipomoea*, such as *I. tricolor* (morning glory), are grown for their showy blooms. To the farmer and gardener the family is better known for its troublesome bindweeds, species of *Calystegia* and of *Convolvulus*.

Calystegia sepium★ **Plate 19**

AKAPOHUE – NAHINAHI – PANAKE – PINK BINDWEED 40–60

A climbing perennial with distinctly triangular or arrow-shaped leaves. The five sepals are enclosed by two large bracts that do not overlap each other. Petals are pink with a white band down the centre, or occasionally all white. The style is longer than the stamens.

It is common on roadsides, waste areas, swamps and forest margins: flowers from September to February.

The name *Calystegia* comes from the Greek *kalyx stegon*, meaning 'calyx cover'; this refers to the two large bracts covering the calyx. The roots and shoots were cooked and eaten by the Maori, and by people in China and India, but in some other parts of the world the plant is regarded as a purgative. Possibly this is due to differences between varieties of the plant, but it is not advisable to use this plant for food.

Calystegia silvatica★ **Plate 19**

GREAT BINDWEED 60–75

A climbing perennial with white flowers, rarely with pink flowers. The leaves are distinctly triangular or arrow-shaped. The sepals are enclosed by two very large bracts that overlap each other. The styles are equal to or only slightly longer than the stamens.

It climbs on hedges and shrubs on the margins of forests and plantations: flowers from October to May.

Calystegia soldanella **Plate 19**

PANAHI – SHORE BINDWEED 30–35

A creeping perennial with pink flowers, each petal having a central white stripe. The leaves are fleshy and kidney-shaped. The sepals are enclosed by two bracts, about the same length as the sepals.

It is found in coastal areas, growing on sand: flowers from October to March.

Stonecrop Family – Crassulaceae

H OWA Si - R
K 3–30 or united
C 3–30 sometimes united
M 3–30 as many or twice as many as petals
F 3–30 sometimes united at the base; as many as petals

As their name suggests, the stonecrops grow in dry, usually rocky, places. Most members of the family have features associated with plants living in such habitats: they have fleshy stems and leaves, and the stems are relatively short with the leaves close together. The leaves usually have a waxy surface.

Several members of the family are cultivated as rock-garden plants or grown to cover walls.

*Cotyledon orbiculata** — Plate 22

COTYLEDON — **30–40 long**

A short, perennial, succulent plant covered with a white bloom. The broad, oval leaves are arranged in loose rosettes, from which grow branching, flowering stems up to 500 mm tall. The flower has five sepals and five petals, the petals being fused into a tube, with spreading lobes.

It grows on dry banks, cliff faces, and similar situations, also sometimes in other dry habitats: flowers from December to June.

*Sedum acre** — Plate 22

BITING STONECROP – WALL PEPPER — **12**

A low, creeping perennial. The leaves are rounded in section, unstalked, not spreading, and have a burning taste (see below).

It grows on walls, banks, cliffs, sand and shingle: flowers from November to March.

This plant is poisonous, so do not swallow the leaf if you try tasting it. The name *Sedum* comes from the Latin for 'home'; this plant was often grown on the roof of a house to protect the occupants from lightning.

*Sedum album** Plate 22

WHITE STONECROP 6–9

A low to short, creeping perennial. The leaves are dark green, brownish green or reddish green, cylindrical, and slightly flattened on the upper surface. They are not as close together as in biting stonecrop and spread out widely from the stem.

Scabious Family – Dipsacaceae

H OB SiC - I
K (5) epicalyx of two fused bracts in some species
C (5)
M 4
F (2)

Some members of this small family (*Scabiosa* spp.) are cultivated as garden flowers. The fuller's teasel (*Dipsacus fullonum*) has an inflorescence with stiff, hooked bracts; it was widely used to raise the nap on woven fabrics, and is still cultivated for this purpose in Somerset, England.

*Dipsacus sylvestris** Plate 13

WILD TEASEL 2

A tall, biennial plant with stiff, spiny stems and leaf veins. The leaves are opposite, with their bases fused (connate) surrounding the stem. The large ovoid inflorescence is surrounded by long, spiny bracts.

It is found beside roads and in waste areas: flowers from October to April.

Water collects in the cup formed by the leaf bases, and this gives rise to its name, from the Greek *dipsa* (thirst).

Sundew Family – Droseraceae

H AB Si – R
K (4, 5 or 8)
C 5
M 4–20 often 5
F 3–5

The most characteristic feature of this family is that the leaves are covered with long, sticky hairs. These bend over when a small insect lands on the leaf, trapping the insect in the sticky liquid secreted from the top of the hair. The secretion contains substances which digest the soft parts of the insect's body; the digested materials are then absorbed into the leaf and used as food materials by the plant. Afterwards the hairs uncurl and the dry outer parts of the dead insect fall away. Since these plants are insectivorous, they are able to live in damp, boggy, waterlogged soils from which the roots of other plants find difficulty in absorbing nutrients. The most notable genus of this family is *Drosera*, of which six species are found in New Zealand. All of these species are native to this country and one (*D. stenopetala*) is endemic.

Drosera arcturi — Plate 23

SUNDEW — **12–16**

A low to short plant with strap-shaped leaves. Long, sticky hairs cover the leaves except near the base. The flower stalks bear one, rarely two or three, flowers. The sepals are very dark green.

It lives in boggy areas from 300 m to 1500 m: flowers from November to March.

The sticky hairs on the leaves give the leaf a dewy appearance, from which we obtain the genus name, derived from the Greek *droseros* (dewy).

Drosera binata — Plate 23

SUNDEW — **15–20**

A short to medium plant with very narrow leaves, forked at least once and covered with sticky hairs. Flower-stalks are dark green or blackish and bear loose clusters of sweet-scented flowers.

It lives in boggy places from lowland up to 300 m: flowers from November to February.

Drosera stenopetala — Plate 23

WAHU — **7–10**

A low to short plant with strap-shaped leaves broadening to a spoon-shaped tip. The tip is edged with prominent, radiating, sticky hairs. The flower-stalks are dark reddish and narrow, bearing a single flower. Sepals are short and rounded.

It lives in boggy areas from lowland up to 1500 m: flowers from November to March.

Heath Family – Ericaceae

S AOW Si – R
K (4–5)
C (4–5) urn-shaped, bell-shaped or saucer-shaped
M 8 or 10
F (4–5)

Members of this family are often low-growing, with leathery, evergreen leaves.

The family includes the genus *Vaccinium* which produces blueberries, bilberries, or cranberries. Several members of the family are grown as ornamental garden plants, including *Rhododendron*, *Pieris*, *Gaultheria* and many varieties of heath (*Erica*).

Erica lusitanica★ — Plate 23

SPANISH HEATH — **3–5 long**

A tall, hairless shrub. Leaves are narrow, with three to four in a whorl. The narrow, bell-shaped flowers, pink in bud, become white when open.

It lives on rocky hillsides and banks, and also in scrub and grassland: flowers from March to December.

Small specimens of tree heath (*E. arborea*★) can be mistaken for Spanish heath, but a distinguishing feature is that the stigma of Spanish heath is red-tipped, while that of tree heath is white or light green. Tree heath grows up to 5 m high.

Gaultheria crassa — Plate 23

KOROPUKA — 5

A slightly hairy, much-branched shrub, up to 1 m tall, though shorter at higher altitudes. Its leaves are alternate, thick, leathery and brownish green; they are about 10–15 mm long and 5–10 mm wide. Flowers are white and waxy, in clusters about 40 mm long. Sepals have reddish tips. Fruits are round, dry capsules, brown when ripe, surrounded by a dry calyx.

Widely occurring from 700 m to 1700 m, it flowers from February to April.

A species of similar appearance is snowberry (*G. depressa*). This is a low-growing shrub with solitary, white flowers; the fruits are fleshy capsules, coloured red, pink or white, surrounded by a dry calyx. Another snowberry belongs to a related genus, *Pernettya*, which often hybridises with species of *Gaultheria*. Plants of *Pernettya macrostigma* are short, prostrate shrubs, with solitary flowers in the axils of the narrow, finely toothed leaves. The fleshy berries are partly surrounded by the fleshy calyx. Berries and calyx are the same colour, from white to dark red.

Spurge Family — Euphorbiaceae

HST AO SiC + R
K 5 sometimes absent
C 5 often absent
A 1–many, free or variously united
G $\underline{(3)}$

Many species in this family show features adapting them to dry conditions. They may have a heath-like or cactus-like appearance, and the leaves and stipules may be reduced to fleshy lobes or spines.

Plants of economic importance include cassava (from which comes tapioca), the rubber tree (*Hevea braziliensis* – many plants of this family produce copious latex) and the castor oil plant. Cassava (or manioc) is especially important as a reserve food plant; the plant is almost immune to attack by locusts and its tuberous roots may be left in the ground unharvested for up to two years without deterioration. The genus *Euphorbia* is interesting because of the variety of forms shown by its species. They vary from the small and delicate garden weed *E. peplus* (see below) to medium-sized and large cactus-like shrubs grown as pot-plants or in shrubbery borders. The genus also includes the well-known 'Christmas plant', *Euphorbia pulcherrima*, otherwise known as poinsettia, which is grown as a pot-plant because of its large scarlet bracts, and is also grown as an ornamental hedge-plant in warmer climates.

Euphorbia helioscopia★ **Plate 23**

SUN SPURGE **1.5**

A medium, erect annual with oval leaves, finely toothed except near the base. The flowers have no sepals and no petals. The inflorescence in this genus consists of a hollow, cup-like structure in which are hidden the male flowers (each consisting of one stamen). The female flower (consisting of a green ovary and style) is on a stalk projecting out of the cup. In this species the inflorescences are in five-rayed umbels (sometimes only three or four rays). Prominent on the margin of the cup are three oval, stalked glands, green in colour. The stem exudes milky latex when broken.

A common weed of gardens and waste places, it flowers all the year round.

Euphorbia peplus★ **Plate 23**

MILKWEED – PETTY SPURGE **1.5**

A low to short annual with oval, entire leaves. The flowers have no sepals and no petals. The inflorescence is as described above for sun spurge, except that the three glands are crescent-shaped, with the points of the crescents directed outward. The inflorescences are in three-rayed umbels. The stem exudes milky latex when broken.

This is a very common weed of gardens, waste places and on sandy or shingly areas: flowers all the year round.

Pea Family – Fabaceae

HSTL A C + I
K (5+)
C 5 or 3+(2) see below
M 10– many see below
F (1)

This family is also known as the Leguminosae, after the typical 'pea-pod' fruit, the legume, or as the Papilionaceae, after the butterfly-like appearance of the flowers of many genera. It is a very large family, containing 12,000 species world-wide. The family is subdivided into three sub-families, of which two contain many trees generally planted as ornamentals in warmer parts of the country:

- Mimosas – tropical and sub-tropical trees such as *Mimosa*, *Acacia* (wattle) and *Albizia* (rain tree).
- Cassias – tropical and sub-tropical trees such as *Cassia*, *Caesalpinia* and *Bauhinia*.

The remainder of the family consists of a wide variety of plants, mainly herbs, as shown by the examples illustrated in this book. The petals comprise a (usually) large standard petal, two wing petals and two keel petals. The keel petals are usually fused along their adjacent edges and enclose the 10 stamens. The filaments of the stamens may be free, or nine of them may be fused to form a tube around the single superior ovary. One stamen remains free so that the tube is not complete.

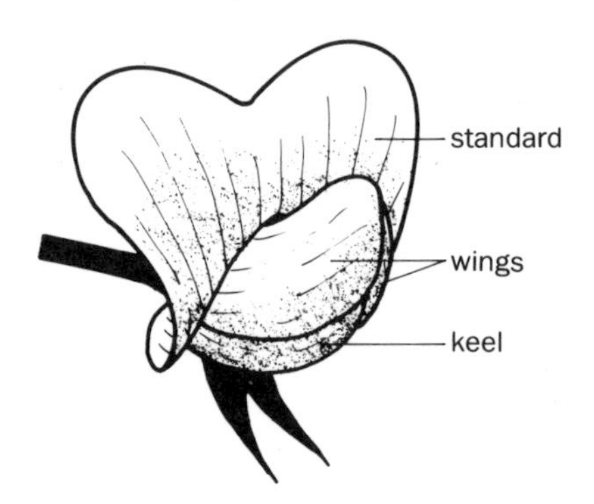

This is a very important family, the second most important source of food after the grasses. One member of the family, the soya bean (*Glycine max*) is particularly productive: it is a source of oil for cooking, and for making margarine, as well as for making soap, paint and plastics. The soya bean sprouts are a major feature of Chinese cuisine. Soya flour has a high protein content and is low in carbohydrate; it is used in making ice cream and soy sauce. Recently, a non-dairy milk substitute has been marketed.

Other economically valuable plants in this family include:

- Seeds used as food: pea, bean (many kinds), lentil, groundnut (or peanut), Moreton Bay chestnut (dried, roasted and eaten by Australian aborigines). All have a high protein content.
- Oil: groundnut
- Dyes: indigo
- Roots: liquorice
- Foliage used as fodder for livestock: clover, lucerne, sainfoin
- Hedge-plant: gorse
- Timber: wattle, rose-wood, Brazil-wood (violin bows)

- Ornamental plants: lupins, sweet pea, brooms, laburnum, *Wisteria*, and the trees already listed above.

A feature of this family is that many of the plants bear nodules on their roots, in which live nitrogen-fixing bacteria; this means that the plants can live and grow on poor soils that are deficient in nitrogenous minerals. Farmers often grow crops of plants such as clover so that, when the roots are ploughed in, the soil is enriched with additional nitrogen.

Lathyrus latifolius★ — Plate 24

EVERLASTING PEA — 20–30 long

A climbing, hairless perennial with winged stems. The stipules are lance-shaped, pointed at either end, usually as wide or wider than the stem. The leaves are parallel-veined, with one pair of leaflets, the remaining leaflets being modified into tendrils. The inflorescence consists of 5–15 flowers. The seeds have a wrinkled surface.

Common in waste places, it flowers from September to May.

Lathyrus tingitanus★ — Plate 24

TANGIER PEA — 20–35 long

A climbing, hairless annual with winged stems. The stipules are lance-shaped, and pointed at either end, usually as wide as the stem. The leaves are parallel-veined, with one pair of leaflets, the remaining leaflets being modified into tendrils. The inflorescence consists of one to three flowers. The seeds have a smooth surface.

It is common in waste places: flowers from August to May.

Lotus pedunculatus★ — Plate 24

LOTUS — 10–13 long

A low to medium perennial with hollow stems and pinnate leaves. There are two pairs of oval leaflets, with distinct side veins. Leaves are unstalked, and there are no stipules, so the lower pair of leaflets can be mistaken for stipules. The inflorescence of 5–12 flowers is on a stalk much longer than the leaves; the flowers are short-stalked, in an umbel-like head. The teeth of the calyx are equal to or less than the length of the calyx tube.

It is found in waste areas and pastures, and in damp habitats: flowers from November to January.

Another plant that resembles this is *L. tenuis**, which has rather narrower leaflets with only the main vein clearly visible, more or less solid stems, and two- to four-flowered inflorescences.

*Lotus suaveolens** **Plate 24**

HAIRY BIRDSFOOT TREFOIL **6–8 long**

A low to medium, shaggily hairy perennial with pinnate leaves. There are two pairs of oval leaflets. Leaves are as in lotus, above. The inflorescence of one to four flowers is on a stalk longer than the leaves, and the flowers are shortly stalked. The teeth of the calyx are longer than the calyx tube.

It is found in waste areas and pastures, and flowers from October to June.

Another shaggily hairy member of this genus is slender birdsfoot trefoil (*L. angustissimus**). The keel is more or less right-angled, contrasting with that of *L. suaveolens*, which is slightly angled and tapers toward its tip, as can be seen in the upper flower at the right of the photograph. The pods of *L. angustissimus* are much longer and narrower.

*Lupinus arboreus** **Plate 26**

TREE LUPIN **15–18 long**

A tall, hairy shrub with palmate leaves, of 5–11 leaflets, 3–10 mm wide. The inflorescence is a raceme of yellow, white or blue-tinged, sweetly scented flowers.

It is common in sandy areas, particularly on coasts: flowers from October to May.

*Lupinus polyphyllus** **Plate 26**

RUSSELL LUPIN **12–20 long**

A tall, hairy, perennial herb, with palmate leaves, of 8–15 leaflets, 10–30 mm wide. Having originated from the cultivated Russell lupins, the flowers have a variety of colours and are often bi-coloured; colours include pink, purple, yellow, white, blue and orange; the flowers are slightly scented or unscented.

It grows in waste areas, commoner in the South Island, and only in Wellington Province in the North Island: flowers from September to February.

A hybrid between the above two lupin species is sometimes found. This has yellow flowers tinged with blue, with wider and longer leaflets and more leaflets in a leaf than the tree lupin.

CLOVERS

Melilotus officinalis
p. 113

Trifolium arvense
p. 114

Trifolium repens
p. 115

Trifolium dubium
p. 114

Trifolium campestre
p. 114

Trifolium pratense
p. 115

Trifolium subterraneum
p. 115

PLATE 26

LUPINS, VETCHES, FUMITORIES

Lupinus arboreus
p. 112

Lupinus polyphyllus
p. 112

Fumaria officinalis
p. 117

Fumaria muralis
p. 117

Vicia hirsuta
p. 116

Vicia sativa
p. 116

Vicia tetrasperma
p. 116

GENTIANS, BELLFLOWERS, FORSTERA

Blackstonia perfoliata
p. 118

Centaurium erythraea
p. 118

Gentianella corymbifera
p. 119

Wahlenbergia albomarginata
p. 93

Gentianella bellidifolia
p. 119

Wahlenbergia gracilis
p. 94

Campanula rapunculoides
p. 93

Forstera bidwillii
p. 170

PLATE 28

GERANIUMS

Geranium purpureum
p. 122

Erodium moschatum
p. 120

Geranium dissectum
p. 121

Geranium pusillum
p. 122

Geranium microphyllum
p. 121

Geranium retrorsum
p. 122

GERANIUMS

Pelargonium ×asperum
p. 123

Geranium robertianum
p. 122

Geranium molle
p. 121

BEDSTRAWS

Sherardia arvensis
p. 161

Galium palustre
p. 160

Galium aparine
p. 160

PLATE 30

MINTS

Ajuga reptans
p. 126

Lamium purpureum
p. 127

Melissa officinalis
p. 127

Prunella vulgaris
p. 128

Stachys arvensis
p. 129

Mentha ×piperita var. *citrata*
p. 127

Mentha pulegium
p. 128

Mentha spicata subsp. *spicata*
p. 128

Basil

Clinopodium vulgare
p. 126

Mallows, stinkweed, periwinkle

Alcea rosea
p. 133

Malva sylvestris
p. 134

Navarretia squarrosa
p. 143

Lavatera arborea
p. 133

Modiola caroliniana
p. 134

Vinca major
p. 55

PLATE 32

LOW-GROWING PLANTS OF DAMP PLACES

Mimulus guttatus
p. 164

Mimulus moschatus
p. 164

Samolus repens
p. 149

Mazus radicans
p. 163

Viola cunninghamii
p. 175

Selliera radicans
p. 124

Pratia angulata
p. 131

Medicago arabica★ **Plate 24**

SPOTTED BUR MEDICK **4–6 long**

A low, prostrate, almost hairless annual, with trefoil leaves, each leaflet having a dark spot in its centre, and an inflorescence of one to four flowers. The pod is coiled into a spiral of three to seven turns, has a double row of spines, and faint net-veining.

It grows in waste places and in grassy places, such as pasture and lawns: flowers from September to April.

A similar species is bur medick (*M. nigra*★), with unspotted leaflets, and distinctly net-veined pods.

Medicago lupulina★ **Plate 24**

BLACK MEDICK **2–3 long**

A low, prostrate, hairy annual, with trefoil leaves, each leaflet ending in a tiny point. It has an inflorescence of 10–15 flowers in a rounded head; the petals fall off immediately after flowering. The pod is coiled in about one turn, and is hairy, but it has no spines.

It grows in waste places, pasture and on coastal sites: flowers from November to May.

Medicago sativa★ **Plate 24**

LUCERNE **7–12 long**

A medium, slightly hairy perennial with trefoil leaves. The inflorescence is a short raceme of violet, blue or sometimes white flowers. The pods are coiled in a tight spiral of up to three turns.

It grows on waste and cultivated areas, and flowers from November to May.

Melilotus officinalis★ **Plate 25**

YELLOW SWEET CLOVER – MELILOT **4–6 long**

A medium to tall, biennial plant. The upper parts are slightly hairy, with trefoil leaves, their margins toothed. The inflorescence is a loose raceme, calyx teeth are shorter than the calyx tube and are triangular, and the keel is shorter than the standard or wings. When crushed, the plant smells like mown grass.

It lives in dry waste areas, particularly on the coast: flowers from November to March.

The name comes from the Greek *meli*, meaning 'bee' and refers to the fact that this plant is particularly attractive to bees.

Parochetus communis★ — Plate 24

SHAMROCK PEA — **18–22 long**

A prostrate perennial with trefoil leaves and flowers of a distinctive blue colour.

It grows in waste places, especially where it is damp and shady: flowers from December to September.

Trifolium arvense★ — Plate 25

HARESFOOT TREFOIL — **3–4 long**

A low to short annual with trefoil leaves, having narrow hairy leaflets. The inflorescence is cylindrical, the flowers being white at first, becoming pinkish later. The sepals are longer than the petals and have long silky hairs.

It is common in waste areas, coastal areas, on pasture and cultivated land: flowers from August to May.

Trifolium campestre★ — Plate 25

HOP TREFOIL — **3.5–5 long**

A low to short, hairy annual with trefoil leaves, with a dark red spot on each leaflet; the middle leaflet has a longer stalk than the other two leaflets. The inflorescence is a broadly rounded head of 20–40 flowers. The flowers are yellow at first, then the petals die and remain, becoming light brown and having a distinctly grooved appearance.

It lives in dry waste areas and in grassland: flowers from November to May.

Trifolium dubium★ — Plate 25

SUCKLING CLOVER — **2.5–4 long**

A low, prostrate annual with trefoil leaves, the leaf tips being slightly notched. The middle leaflet has a longer stalk than the other two leaflets. The inflorescence is a rounded head of only 5–20 flowers, and the petals turn brown when dead, just covering the straight pods.

It lives in grassy places, including lawns and playing-fields, roadsides and waste ground: flowers from October to June.

Suckling clover is the commonest of the yellow clovers. At first glance it resembles some of the smaller, yellow-flowered oxalis species, but in suckling clover there are several flowers in a head, while in the oxalis the flowers are not grouped in heads.

Trifolium pratense★ — Plate 25

RED CLOVER — **10–16 long**

A short to medium perennial with hairy, trefoil leaves, the leaflets usually with a whitish crescent-shaped mark. The inflorescence is terminal, rounded or ovoid, many-flowered, and very shortly stalked, with two leaves close beneath. The sepals are hairy, usually half as long as the petals. Occasionally the flowers are cream or white.

It is common in pasture, cultivated and waste land: flowers from October to March.

The name 'clover' comes from the Latin *clava*, meaning a 'club'; the symbol on the clubs suit of playing cards represents the clover leaf. The phrase 'living in clover' refers to the contentedness of cattle grazing a field of lush clover plants.

Trifolium repens★ — Plate 25

WHITE CLOVER — **8–15 long**

This is a creeping, perennial plant, the stems rooting near the bases of the leaves. The plant is almost completely hairless, with possibly a few hairs on the leaves and calyx. It has trefoil leaves, usually with a whitish crescent on each heart-shaped leaflet, and straight side veins, remaining thin at the leaf margins. Inflorescences are produced in the axils of leaves, on long stalks; they are rounded and many-flowered.

It grows in a wide variety of places: flowers from July to March.

The strawberry clover (*T. fragiferum*★) has the same creeping and rooting habit, but has pale pink flowers. The flower-heads swell on fruiting to resemble pale pink strawberries. Its leaves are narrower and have curved side veins that thicken toward the leaf margins.

Trifolium subterraneum★ — Plate 25

SUBCLOVER — **10–15 long**

A short, prostrate plant with hairy trefoil leaves and inflorescences of few (two to five) large, fertile flowers; smaller, sterile flowers develop later. As the fruits ripen, the flower-stalks curl downward, burying the ripening pods in the soil.

It grows in waste and grassy places: flowers from September to February.

The ripening pods are turned back as the flower-head touches the soil. At the same time the calyxes of the sterile flowers enlarge, anchoring the seed-pods in the soil.

Vicia hirsuta★ — Plate 26

HAIRY VETCH — **4–5 long**

A short annual with pinnate leaves ending in tendrils. The leaves have five to nine pairs of narrow leaflets. Stipules are unspotted. The inflorescence is a raceme of three to seven flowers, on a long, leafless stalk. The calyx teeth are more or less equal and equal to or longer than the tube. Flowers are dull white or pale blue.

It lives in waste and cultivated places, inland or on the coast: flowers from September to April.

In spite of its common name, the stems and leaves of the hairy vetch may be hairless or only slightly hairy. Its name presumably comes from its hairy pods.

Vicia sativa★ — Plate 26

COMMON VETCH — **10–30 long**

A short to medium annual with pinnate leaves ending in tendrils. The leaves have four to eight pairs of oval to linear leaflets. There is usually a dark spot on the stipules. Flowers are one or two on very short stalks in the axils of the leaves. The calyx teeth are more or less equal, and may be longer than, the same length as, or shorter than the tube. Flowers may also be reddish purple, rose, pink or white.

It lives in waste and cultivated places, inland or on the coast: flowers from August to June.

Vicia tetrasperma★ — Plate 26

SMOOTH TARE — **4–8 long**

A short to medium annual with slender, angular stems and pinnate leaves ending in tendrils. There are two to five pairs of narrowly oval leaflets, which may be arranged alternately on the leaf stalk, as in the photograph. It has an inflorescence of one to three flowers on a long stalk. The pods have four (rarely five) seeds.

It lives in waste and cultivated places: flowers from October to May.

Fumitory Family – Fumariaceae

H A C – I
K 2 soon falling off
C 2+2 dissimilar
M 2
F (2)

These are low-growing plants with weak stems. The flowers are two-lipped, with spurs at the bases of the petals.

*Fumaria muralis** — Plate 26

WALL FUMITORY — **8–10 long**

A short, annual plant with racemes of up to 15 flowers, the stalk of the raceme being as long as or longer than the raceme itself. The flowers are larger than those of fumitory. The lower petal is spoon-shaped, but its edges are turned up at the tip. The tips of the petals are dark purple. The fruits are slightly longer than broad, smooth when dry.

It is common in cultivated and waste areas: flowers from September to March.

*Fumaria officinalis** — Plate 26

FUMITORY — **7–8 long**

A short, annual plant usually with a variable number of flowers in the raceme, the stalk of the raceme being rather less than the length of the raceme. The flowers are smaller than those of wall fumitory, with the lower petal being spoon-shaped and having a wide-spreading margin at the tip. The tips of the petals are purple and dark green. The fruits are slightly broader than long, rough when dry.

It grows in cultivated and waste places, but is less common than wall fumitory: flowers from October to May.

One of the common names of this plant was 'earth smoke' and the word 'fumitory' is derived from this, via the Latin *fumus terrae*. It is thought that the plant was given this name because, when seen at a distance, the greyish leaves look like smoke rising from the ground.

Gentian Family — Gentianaceae

HfewS O E – R
K (4–5)
C (4–5) bell-shaped, funnel-shaped or saucer-shaped
M 4–5 alternate
F (2)

Most members of the family are hairless herbs, with unstalked leaves.

A few plants of this family are grown as ornamental garden plants or house plants, but, on the whole, this family is not of economic importance.

Blackstonia perfoliata★ — Plate 27

YELLOW WORT — **15–20**

A short to medium annual. The stem and leaves are a greyish green, the leaves are wide at the base and the bases of opposite leaves are fused, surrounding the stem. The flowers have six to eight petals.

It is found only in the North Island, where it grows on roadsides and beaches and in pasture and scrub: flowers from December to February.

Centaurium erythraea★ — Plate 27

CENTAURY — **10–12**

A low to short, hairless annual or biennial, growing from a basal rosette, with a few narrow, three-veined leaves on the stem. The bracts of the inflorescence are longer than the sepals of the flowers. Flowers are almost unstalked in dense clusters.

It is a very common plant of open habitats from lowland to 700 m: flowers from November to April.

A plant of similar appearance is *C. tenuiflorum*★; but this does not have a well-defined basal rosette at flowering time. The flowers are shortly stalked, and smaller, with narrower, more sharply pointed corolla lobes.

The gentians

These showy plants are some of the more conspicuous members of the New Zealand alpine flora. There are 24 native species of the genus, which were formerly included in the genus *Gentiana*. Recently, many of these have been transferred to two other genera, *Gentianella* and *Oreophylax*; here we refer to them as *Gentianella*. Since many of the gentians are confined to the South Island and to mountain areas, we restrict ourselves to describing two gentians, as typical examples of the group.

Gentianella bellidifolia — Plate 27

GENTIAN — **20**

A short perennial with fleshy, tufted basal leaves, oval to spoon-shaped, 10–15 mm long, 5–7 mm wide. There are several stems, unbranched or sparingly branched, and three to four pairs of smaller stem leaves. The flowers are terminal and solitary or found occasionally in clusters of up to six. The calyx is about half the length of the corolla; its lobes are at least three times as long as its tube and the gap between the margins of adjacent lobes is narrow. The fine, purple veins on the petals are a distinguishing feature.

It lives in tussock grasslands and boggy places from 600 m to 1800 m: flowers from January to March.

Gentianella corymbifera — Plate 27

GENTIAN — **12–20**

A medium perennial with fleshy, tufted basal leaves, more or less lance-shaped, 50–150 mm long, 6–20 mm wide. The stem is single, usually unbranched, with a few pairs of smaller, lance-shaped leaves, and ending in a dense cluster of flowers. The calyx is about half the length of the corolla. Its lobes are about equal in length to the tube, or slightly shorter.

The gentian lives in tussock grasslands in the South Island, from 400 m to 1500 m: flowers from January to March.

Geranium Family – Geraniaceae

H AO Si + IR
K 5 sometimes fused half-way
C 5
M 5, 10 or 15 sometimes united at the base
F (3–5)

Most of the family are hairy herbs, and most fruits end in a long, narrow 'beak', leading to the names 'storksbill' and 'cranesbill' for several members of the family. When the fruit is ripe and dry, the carpels split away at the base and curl upward, still attached to the central beak; in some species this happens so violently that the seed is thrown from the plant. In other species the whole carpel wall, and the enclosed seed, suddenly splits away and travels some distance from the plant. On landing in moist surroundings, the curled portion of the carpel becomes straighter and this action often tends to force the seed down into the soil.

The best known genus is *Pelargonium*, of which very many colourful or highly scented varieties are grown in gardens and as house plants. These are generally, though incorrectly, referred to as 'geraniums'.

*Erodium moschatum** — Plate 28

MUSKY STORKSBILL — 8–12

A low to medium, glandular-hairy plant, sticky to the touch, with a musky smell. The leaves are pinnate, up to 300 mm long, with oval, toothed leaflets, rarely pinnately lobed. The inflorescence is an umbel of about 12 flowers, with 10 stamens in two whorls of five, the inner whorl without anthers. The beak of the fruit is 30–35 mm long, covered with very fine, very short hairs.

It lives in a variety of open places, including roadsides, lawns and pasture: flowers from September to November.

Storksbill (*E. cicutarium**) lacks the musky smell of musky storksbill and is less sticky. Its leaves are finely divided and fern-like. In dry habitats or on poor soil, plants of both species may be red in colour and stunted.

Geranium dissectum★ — Plate 28

CUT-LEAVED CRANESBILL — **8–10**

A short to medium, hairy annual, with the hairs all backwardly-pointing. The rounded leaves are cut almost to the base, into five to seven segments, which may be three-lobed or pinnately lobed. The flowers are usually in twos, the petals notched, with a short claw. The stalk of the inflorescence is shorter than the leaf-like bract at its base. The sepals spread out widely and end in an awn 1 mm or more long.

It lives in bare places, especially around buildings: flowers from November to February.

The roots of this plant are edible. The roots of the related matuakumara (*G. homeanum*), growing as a weed in the kumara beds, were eaten by the Maori when food was scarce.

Geranium microphyllum — Plate 28

CRANESBILL — **15–20**

A low, softly hairy perennial, the hairs on the stems backwardly-pointing. The leaves are rounded to kidney-shaped in outline, but three- to five-lobed, usually almost to the midrib. The flowers are usually in twos, the petals slightly notched, with a short claw. The sepals spread out widely and end in a short awn. The beak of the fruit is 7–11 mm long.

It lives in grassland among tussocks and in rocky places from lowland to 1500 m: flowers from October to February.

Geranium molle★ — Plate 29

DOVESFOOT CRANESBILL — **8–12**

A low to short annual with rounded leaves cut to about three-quarters way, into five to nine segments. The stem is softly hairy, especially near the base. The flowers are usually in twos, and droop after flowering. Petals are deeply notched, with a very short claw. The sepals spread out widely, ending in a sharp tip. The fruit is hairless.

It is common in waste and cultivated places, and flowers from September to February.

Geranium purpureum★ **Plate 28**

SMALL HERB ROBERT – LITTLE ROBIN **10**

A medium, hairy annual or biennial with reddish stems and leaf-stalks. The leaves may also be reddish and are divided into five lobes, which are in turn pinnately lobed. Flowers are in twos with slightly notched petals, the claw being slightly longer than the limb; sepals are reddish, and upright; the anthers produce yellow pollen.

It is found in the North Island, growing on roadsides and waste areas: flowers from September to January.

Geranium pusillum★ **Plate 28**

SMALL-FLOWERED CRANESBILL **5–7**

A low to short, densely hairy annual with rounded leaves, cut more than half-way into five to nine segments. The segments are divided into two to three narrow lobes. The stem hairs are short, and the stem is only slightly hairy at the base. Flowers are often in twos with slightly notched petals, which have a short claw. The sepals are spread widely and end in a prominent awn. Many of the stamens have no anthers.

It grows in waste areas, and flowers from October to February.

Geranium retrorsum **Plate 28**

CRANESBILL **6–10**

A medium perennial with rounded leaves cut almost to the base into lobed segments. The hairs are short and bent near the base, pointing backward down the stem and leaf-stalks, and are closely pressed to them. The flowers are in twos. The sepals spread widely and have a prominent awn about 1 mm long.

It lives in grassland and scrub: flowers October to April.

Geranium robertianum★ **Plate 29**

HERB ROBERT **20**

A medium annual with leaves of three leaflets, one- to two-pinnately lobed. The stems and leaves are usually tinged in red or are almost entirely red. The plant has a strong and rather unpleasant smell. Flowers are in twos; the petals, with three whitish stripes, and are very slightly notched at the tips. The limb and claw are equal in length. Sepals are erect and end in an awn. The pollen is orange.

It is widely distributed on roadsides, in gardens, scrub and forest margins, especially in shady locations: flowers from September to May.

There are many theories about the naming of this plant. One is that the name 'geranium' derives from the Greek *garanos*, meaning 'crane'. The name 'Robert' is also thought to come from Greek, from *ruberta*, meaning 'red' and referring to the distinctive red colour of the stems and leaves. However, other authorities believe that the plant was named after Robert Duke of Normandy, a famous medical scholar. Another candidate for having the plant named after him is Abbé Robert, who founded the Cistercian Order in the 11th century.

Pelargonium ×asperum★ **Plate 29**

GERANIUM **15–20**

A medium to tall, hairy, strongly aromatic sub-shrub. The leaves are triangular, deeply cut into three lobes with toothed, wavy margins. The side lobes each have a sub-lobe, while the terminal lobe has two sub-lobes. The inflorescence is an umbel of about 10 flowers on a relatively long stalk. The upper sepal has a spur, joined to the flower-stalk. The two upper petals are larger than the three lower petals and have distinctive markings. The style and stigmas are rose-coloured.

This is a common garden escape and is most often found in coastal areas: flowers throughout the year.

Several other hybrids are widely cultivated in gardens and have escaped. Nutmeg geranium (*P. ×fragrans*★) is a bushy plant smelling of nutmeg; its upper two petals have crimson markings. Regal pelargonium (*P. ×domesticum*★) has triangular to broadly oval leaves, and purplish or dark crimson markings on the upper petals, including a central patch. Zonal pelargonium (*P. ×hortatum*★) is recognised by the dark ring on its leaves; its flowers are bright red. The native pelargonium *P. inodorum* is a non-aromatic, scrambling, hairy plant with rounded, usually unlobed leaves and smaller, pink or rose flowers. It has purple streaks on the upper two petals.

Selliera Family — Goodeniaceae

HS A Si – I
K 5
C (5) tube may be split on one side
M 5 or (5) alternate
F 1 stigma surrounded by a pollen cup

Most members of this family are natives of Australia. The family is closely related to the Lobelia Family, which also has several members with split corolla tubes.

Selliera radicans **Plate 32**

SELLIERA **7–10**

A low, creeping and rooting, mat-forming perennial. The leaves are yellowish green and fleshy, strap-shaped or spoon-shaped, up to 40 mm long. The flowers are solitary on stalks about as long as the leaves. The corolla tube is split to the base along one side. The anthers are free, not fused into a tube around the style. The pollen cup surrounding the stigma is a distinctive feature.

It grows in salt marshes and other muddy or sandy places where water is brackish or that are subject to salt spray: flowers from November to April.

Raspweed Family — Haloragaceae

HS AO Si – R
K (2–4)
C 2–4
M 4–8 usually twice as many as petals
F (2–4)

Several members of this family are aquatic, including water milfoil (*Myriophyllum* sp.), frequently grown in aquaria and ornamental garden pools.

Haloragis erecta **Plate 17**

RASPWEED **2**

A medium to tall, perennial plant with opposite, stalked leaves. The small flowers occur in clusters of three to seven, and have eight stamens.

It grows on roadsides, and waste places: flowers from December to February.

The plant in the photograph belongs to the common subspecies *erecta*. The other subspecies *cartilaginea* has more rounded, thicker leaves, and a relatively shorter leaf-stalk.

Hectorella Family — Hectorellaceae

S A E - R
K 2
C 5
M 5 alternate
F 1

A family consisting of only two genera with only one species each.

Hectorella caespitosa **Plate 38**

HECTORELLA **7–8**

A low, cushion-forming sub-shrub. The leaves, which are fleshy and hairless, are spirally arranged in tight rosettes at the ends of the branches. When in flower, the cushion is almost covered with blooms.

It lives in the South Island, in fellfields from 1300 m to 2000 m: flowers in January and February.

Similar cushion-plants with small white flowers are shown on plate 38.

Mint Family – Lamiaceae

HS O Si - I
K (5) bell-shaped or funnel-shaped tube, may be two-lipped
C (5) usually two-lipped
M 2+2 sometimes two only
F $\underline{(2)}$ see below

Known also as the Labiatae, this family has the characteristic feature of square stems. Adjacent pairs of leaves are arranged at right-angles; they are usually hairy and often have glands producing aromatic oils. As the carpels develop they each become divided into two, so that the developing fruit is divided into four, and looks like a miniature hot cross bun.

The family is well known for its aromatic smells:

- Oils and perfumes: lavender, rosemary, patchouli.
- Culinary herbs: rosemary, mint, peppermint, spearmint, sage, thyme, lemon thyme, summer savory, winter savory, balm, basil, marjoram.

Species of *Salvia*, as well as thyme and bergamot, are often grown as ornamental plants.

*Ajuga reptans** — Plate 30

BUGLE — **13–17 long**

A low to short perennial with rooting runners. The stem is hairy on two opposite sides, and the leaves are only slightly toothed and often have a purplish, 'metallic' appearance (cultivated variety 'Atropurpurea'). The flower has a very small upper lip, and the lower lip has three distinct lobes, the central one being notched.

A garden escape, surviving in shady places, it flowers all the year round.

*Clinopodium vulgare** — Plate 31

WILD BASIL — **7–9 long**

A short to medium perennial, with long hairs on stems and leaves. It has up to four dense whorls of 10–20 flowers. The bracts of the inflorescences are bristle-like and white-haired. The calyx is two-lipped, with a curved tube, usually with 13 veins, often purplish, and covered with long white hairs.

It lives in damp places by roadsides, in scrub and forest clearings: flowers from December to May.

Lamium purpureum★ — Plate 30

RED DEAD NETTLE — **10–14 long**

A low to medium, downy annual, with broad, heart-shaped, bluntly toothed, stalked leaves. There is a ring of hairs inside the corolla tube, near the base.

It grows on waste and cultivated land: flowers from September to November.

Cut-leaved dead nettle (*L. hybridum*★) is similar to red dead nettle but its leaves are irregularly and more deeply toothed, and it lacks the ring of hairs inside the corolla tube. Its leaves are broad and rounded, often almost kidney-shaped. The lower lip of the corolla has a large middle lobe, divided into two, and two small side lobes. It flowers from June to December. Henbit (*L. amplexicaule*★) is so called because the coarsely rounded teeth make the long-stalked leaves look as if they have been bitten by hens. Its distinctive feature is that leaf-like bracts in the inflorescence are unstalked and the bases of opposite bracts are fused together, completely surrounding the inflorescence stalk. It flowers from August to January, but the flowers often do not open, as the plant is self-pollinated.

Melissa officinalis★ — Plate 30

LEMON BALM — **12–15 long**

A medium, hairy perennial, smelling of lemon. There are few flowers in bloom at one time, the whorls consisting mainly of green sepal tubes, 13-nerved, with long, white, shaggy hairs. The tube of the corolla is distinctly curved.

It lives on roadsides, waste areas, and forest margins and scrub: flowers from December to May.

Its name comes from the Greek word for bee, *melissa*. This refers to the attractiveness of this plant to bees.

Mentha ×*piperita* var. *citrata* — Plate 30

PEPPERMINT — **5–6 long**

A short to medium, hairy perennial with a pleasant, lemony smell. The leaves have distinct stalks and sharply pointed teeth. The inflorescence is dense and elongated with a rounded head and a few rounded clusters of flowers below.

It lives in and around streams, rivers and lakes: flowers from January to May.

This is a hybrid of water mint *Mentha aquatica* and spearmint *Mentha spicata* (see below). Variety *citrata* is sometimes known as bergamot mint. The other variety is *piperita**, which has a peppermint smell, narrower leaves and an elongated inflorescence, rarely interrupted; the Romans used to flavour their wine with this mint. In those days, wine was only for gods and men, and women were liable to the death penalty if caught drinking it. Many women made a paste of mint and honey, which they chewed to mask the smell of wine that they had drunk in secret.

*Mentha pulegium** — Plate 30

PENNYROYAL — **4–6 long**

A short perennial, with a strong smell. Leaves are oval with entire or slightly toothed margins. The inflorescence has no terminal head, and consists of separated, dense whorls of flowers.

It lives in pastures and beside rivers and lakes: flowers from November to May.

In ancient times the plants were strewn on the floor to keep away mice and fleas (Latin *pulices*). It was also used as a stuffing for black pudding in the north of England, where it was known as 'pudding grass'.

Mentha spicata subsp. *spicata** — Plate 30

SPEARMINT — **3–5 long**

A short to medium, hairless perennial with a 'mint sauce' smell. Leaves are lance-shaped or slightly broader, with no stalks or very short stalks. The inflorescence is a narrow, pointed spike, consisting of dense clusters of flowers.

Subspecies *tomentosa* has long, soft, grey hairs on the stems and leaves. The plant was originally known as 'spire mint' because of the shape of its narrow, pointed spikes.

*Prunella vulgaris** — Plate 30

SELF-HEAL — **10–15 long**

A low perennial, with no smell. The leaves are entire or with shallow teeth. The inflorescence is dense and parallel-sided, with two leaves immediately below the head. The upper lip of the flower is helmet-shaped, and the calyx has two lips, about half as long as the corolla. The corolla is two-lipped, the upper lip hooded, the lower lip three-lobed, with small teeth on the edge of the middle lobe.

It lives in damp paddocks, lawns and waste land: flowers from November to April.

At one time this was used as a cure for sore throats, a condition known as 'brunella'. The more recent name 'prunella' is derived from this.

*Stachys arvensis** — Plate 30

STAGGERWEED — **6–7 long**

A low to medium, hairy annual with no smell. The leaves are oval or rounded, with round-toothed margins, and the upper leaves are unstalked. The corolla is only a little longer than the calyx; the calyx teeth are about as long as the tube. The upper lip of the corolla is small and not hooded, shorter than the lower lip.

It grows in waste areas, gardens, paddocks and beside roads: flowers throughout the year.

Hedge woundwort (*S. sylvatica**) is named because of its blood-red corolla with white markings. It has a strong, unpleasant smell, and is much taller than staggerweed. It is common in the North Island, but is relatively uncommon in the South Island.

Flax Family – Linaceae

HfewS AO E +– R
K 5 sometimes four sepals, sometimes slightly fused at the base
C 5 sometimes four petals
M 5 or 10 united at the base
F (2, 3 or 5)

Most plants of this family have long, slender, relatively unbranched stems, with small leaves. They are hairless.

The most important plant in this family is linseed (*Linum usitatissimum*, see below); the name *usitatissimum* means 'most useful' and is well deserved, though the plant is less often used nowadays than formerly. The stems of this plant are the source of flax fibre, which is made into linen fabric and thread, while the seeds are crushed to yield linseed oil, used as a base for paints and for treating timber. At one time it was widely used for making the floor covering linoleum, or 'lino'. Several species of *Linum* are grown as ornamental plants.

*Linum bienne** — Plate 33

PALE FLAX — **20**

A short to tall, hairless annual or perennial, with small, narrow leaves alternately arranged on upright, usually slender, stems, which may branch near the base. Petals are 7–11 mm long, sepals 4–6 mm long.

It grows in waste areas and pastures: flowers from November to April.

Linseed (*L. usitatissimum**) has larger flowers (petals 12–18 mm long, sepals 6–10 mm long) and leaves, and may have white petals.

*Linum trigynum** — Plate 33

YELLOW FLAX — **12**

A short to medium annual similar to pale flax, but with smaller, yellow flowers.

It grows in waste places and in sandy areas near the coast: flowers from September to July.

Lobelia Family – Lobeliaceae

HS A Si – I
K (5)
C (5) tube split open on one side in some species
M (5) anthers form a tube round the style
F (2–3)

The most notable members of this family are the giant tree lobelias of the East African mountains. There are several species of these, each of which is usually restricted to a particular mountain, and they grow to about 4 m high, with leaves up to 250 mm long. By contrast, the more familiar members of the family are small herbs.

Lobelia anceps — Plate 21

LOBELIA — **6–10 long**

A low, spreading, hairless herb with narrowly winged stems. The leaves vary in shape and thickness, and may be quite fleshy, like those in the photograph, or thin like a more typical leaf. The lower leaves are broader and more rounded than the upper leaves, which are narrow and strap-shaped. The lower leaves have flat leaf-stalks, while the upper leaves are stalkless. The margins of the leaves are entire or slightly but distantly toothed, and they run down into the wings on the stem; in some plants all leaves are alike. The flowers are on short stalks in the axils of the upper narrow leaves, and may be white, pale blue or pale pink. The corolla tube is split to the base along one side.

It lives mainly in coastal areas on sand and rocks: flowers from September to May.

*Lobelia erinus** — Plate 21

BEDDING LOBELIA — **11–17 long**

A short, annual plant, usually with trailing stems. The leaves are variable, sometimes broader than those shown in the photograph, and occasionally spoon-shaped; margins may be toothed or entire. The corolla tube is split to the base on the upper side.

A garden escape growing on roadsides and other open sites close to gardens: flowers from September to May.

This plant was the subject of experiments by Charles Darwin. He found that bees did not visit the flowers if the blue petals were cut off, which showed that bees locate the flowers by sight, not by smell.

Pratia angulata — Plate 32

PANAKENAKE — **7–12**

A low, hairless or almost hairless plant, with spreading, rooting stems, forming a mat. The rounded leaves (2–12 mm long, 2–8 mm wide) have entire or coarsely toothed margins. Flower-stalks are up to 60 mm long, and much longer than the leaves. The corolla tube is split to the base on the upper side.

It lives in damp situations beside streams, under light trees, in grassland and occasionally on lawns: flowers from October to April.

The leaves were cooked and eaten by the Maori.

Loosestrife Family – Lythraceae

HfewST O E – R
K 4–8 epicalyx of fused stipules; calyx tube (see below)
C 4–8
M 8–16 or more, usually twice as many as sepals
F (2–6)

The bases of the sepals are often fused into an elongated calyx tube; the stamens are mounted at the lower end of this and the petals are mounted on its rim.

The red dye henna is obtained from *Lawsonia inermis.*

*Lythrum hyssopifolia** — Plate 33

HYSSOP LOOSESTRIFE — 4–6

A short to medium annual with pinkish, square stems. Leaves are narrow and alternate, and the flowers are small, with six pink, crumpled petals and four to seven stamens of different lengths.

It grows in open damp places, including roadsides: flowers from December to February.

*Lythrum junceum** — Plate 33

ROSE LOOSESTRIFE — 8–12

A short to medium annual or perennial with pinkish, square stems. The leaves are narrow and alternate. The flowers are larger than those of hyssop loosestrife, with six rose-pink, crumpled petals. The calyx tube has reddish spots half-way along it.

It lives on damp ground beside inland waters: flowers from December to February.

Mallow Family – Malvaceae

HST A Si + R
K (5) may be an epicalyx of bracts
C 5
M many see below
F (1–5–many)

The leaves of the Mallow Family are often palmately lobed. The stamens are often branched, their filaments being united into a tube surrounding the style and fused to the base of the petals.

Cotton (*Gossypium*) is the most important plant of this family; the seeds of this genus have a mass of long hairs, from which cotton is spun. Another economically important plant is okra (ladies' fingers), a species of *Hibiscus* widely grown as a vegetable in the tropics and sub-tropics. Several other species of *Hibiscus* are cultivated in warmer regions as ornamental shrubs and trees. Other ornamental plants in this family include mallow, *Lavatera*, *Sidalcea* and hollyhock.

*Alcea rosea** — Plate 31

HOLLYHOCK — **40–80**

A tall biennial or perennial with an unbranched, erect stem. The leaves are rounded, not lobed or lobed only shallowly, and the flowers are solitary, or in clusters of two or three in the axils of the leaves. They may be white, cream, or various pinks, reds and purples.

It lives in dry waste areas in the South Island.

*Lavatera arborea** — Plate 31

TREE MALLOW — **15–20 long**

A tall, stout biennial with velvety-hairy, rounded, lobed leaves; leaf margins are bluntly toothed. The epicalyx has three segments, longer than the five calyx segments, as can be seen in the flower buds in the upper half of the photograph. At fruiting, the epicalyx segments enlarge and spread widely, while the calyx segment curls inward.

It is found in coastal areas, on cultivated and waste land: flowers from August to May.

*Malva sylvestris** — Plate 31

LARGE-FLOWERED MALLOW — **25–40**

A medium to tall, sparsely hairy perennial. The leaves are palmately lobed, cut less than two-thirds to the base, the lobes bluntly toothed. The flowers are in small clusters of two or more in the axils of the leaves; petals are two to four times as long as the sepals.

It grows in waste places, roadsides, gardens and arable land. Less common than other mallows, but more noticeable, it flowers from November to April.

Small-flowered mallow (*M. parviflora**) has petals 3–5 mm long, only slightly longer than the sepals. Dwarf mallow (*M. neglecta**) is a prostrate annual with white to lilac flowers, the petals 8–15 mm long, about twice as long as the sepals.

*Modiola caroliniana** — Plate 31

CREEPING MALLOW — **10**

A low, creeping annual with broad leaves deeply and palmately divided. The orange to red flowers are conspicuous in spite of their relatively small size.

It grows in waste and coastal areas: flowers from October to May.

Willowherb Family – Onagraceae

HfewST AOW Si – R
K 4 calyx tube, see below
C 4
M 4 or 8
F (4)

In this family the receptacle is usually extended to form a saucer-shaped calyx tube at the end of the ovary; the other flower-parts are mounted on this.

There are no important food plants, though two of minor importance are water chestnut and pomegranate. The family includes a number of garden plants, such as fuchsia, evening primrose, *Clarkia* and *Godetia*. New Zealand willowherb (*Epilobium nerterioides*) has the distinction of being one of the few New Zealand native plants that has successfully established itself in Britain; it was first recorded near Edinburgh in 1904, and is now well established throughout the north-west of Britain and in Ireland.

Epilobium billardiereanum — Plate 33

WILLOWHERB — **10–20**

A tall, perennial, erect plant. Leaves are narrow, or narrowly oval, toothed, with very short stalks. Flowers may be white, as in the photograph, or rose-purple; pods are greyish green.

It lives in a wide range of habitats from sea-level to about 300 m: flowers from December to February.

Epilobium glabellum — Plate 33

WILLOWHERB — **10–15**

A medium, bushy perennial with red-tinged stems and leaves. The plant is rather variable: the leaves may be lance-shaped or narrowly oval, with shallowly toothed margins; the flowers can be white or rose-purple, and they may be tinged pink after pollination.

It lives on damp and stony ground from lowland to 2500 m in mountain areas: flowers from November to March.

The willowherbs

A total of 42 species of willowherb can be found in New Zealand, 37 of them native. Many of the native species are mountain plants, and relatively uncommon. Identifying individual willowherb species is difficult, but the *Epilobium* genus has features that make it easy to pick out a 'willowherb' from other types of plant; the photographs on plate 33 show species that are typical of the group as a whole. In particular, note:

- Narrowly oval leaves (reminding us of willow leaves).
- Petals are alike and notched.
- Petals are white or a shade of pink.
- The fruit is a very long and narrow pod (from the inferior ovary).
- The walls of the fruit splits into four narrow segments.
- The pod contains numerous seeds with a tuft of silky hairs.

Epilobium tetragonum★ **Plate 33**

TALL WILLOWHERB **3–5**

A short to tall perennial with stems branching at the base to form dense clumps. Leaves are narrowly lance-shaped, with finely and shallowly toothed margins; and they are stalkless, their margins running down into raised lines on the stem. Flowers are rose-purple to mauve.

Common in northern areas of the North Island, it grows in damp places such as ditches and beside streams: flowers from November to January.

Oenothera glazioviana★ **Plate 43**

EVENING PRIMROSE **60–80**

A tall, hairy biennial with the stem covered with dark red hairs with swollen bases. The flowers open in late afternoon or evening. The fruits are more or less cylindrical, but widening toward the base.

It is found in waste areas and on roadsides: flowers from December to April.

The flowers of this plant are very short-lasting; perhaps it was this fact that led young Victorian men and women to send the evening primrose to lovers they suspected of being inconstant.

Oenothera stricta★ **Plate 43**

SAND PRIMROSE **40–90**

A medium to tall annual or biennial, it has narrow leaves, with finely and distantly toothed margins; the leaves are hairless except at the edges. The older petals become reddish orange, as on the left of the photograph. It is fragrant in the evenings. The fruits are more or less cylindrical, but narrow toward the base.

Broomrape Family – Orobanchaceae

H A E – I
K (4) **or** (5) two-lipped
C (5) two-lipped
M 4
F $\underline{1}$

All members of this family are parasitic on the roots of other plants. They lack the green pigment chlorophyll, so they may be whitish or variously coloured in light brown or purple. Since they have no chlorophyll, they do not produce food materials by photosynthesis, and their leaves are reduced to scales. The stem ends in an inflorescence consisting of scaly bracts and flowers, both of which are usually the same colour as the stem and leaf-scales.

*Orobanche minor** — Plate 41

BROOMRAPE — **10–18 long**

A short to medium annual, with a straight, unbranched stem bearing yellowish scale-leaves and a spike of pale yellowish flowers. The plants may also be coloured purplish.
It is found as a parasite on plants of pastures and crops, on roadsides and in waste areas: flowers from August to January.

It produces large quantities of dust-like seeds, which account for its success as a weed.

Wood Sorrel Family – Oxalidaceae

HfewS A C + R
K 5 **or** (5) often have orange swellings, or calli
C 5 **or** (5)
M 5+5 outer whorl opposite petals, rarely 15
F $\underline{5}$ **or** $\underline{(5)}$

The family is best known for the members of one genus, *Oxalis*, of which many species and varieties are ornamental plants. The tubers of *Oxalis tuberosa*, known as oca, have young shoots that are eaten as a salad vegetable.

Oxalis articulata★ **Plate 37**

SOURGRASS **20**

A low to short, hairy perennial with leaves in basal rosettes. There are orange spots (calli) on the underside of the leaves, and on the ends of the sepals. Leaflets are hairy above and below, unstalked, often unequal in size, divided by a narrow cut into two rounded lobes. The flowers are in clusters of up to 35; flower stalks are curved back when the flowers are in bud, but become erect as the flowers open. The ends of the petals are cut across at an oblique angle. Petals are rose, with darker veins, pale pink on the outside.

This is a garden escape, which has become established on roadsides and waste areas: flowers from July to May.

Two similar plants with pink flowers are pink shamrock (*O. debilis*★) and fishtail oxalis (*O. latifolia*★). Both produce numerous small bulbs (bulbils) around a wide, contractile root and, for this reason, are troublesome weeds. Sourgrass does not have a contractile root or bulbils. In pink shamrock the leaflets are hairy below but not or scarcely hairy below, and have hairy stalks about 1 mm long. In fishtail oxalis the leaflets are divided by a broad cut into two tapering segments, giving the leaflet a distinctive fishtail outline.

Oxalis corniculata★ **Plate 37**

HORNED OXALIS **4–7**

A low to short perennial, with creeping, rooting stems. Leaves are alternate along the stem, and without calli. The leaflets are very shortly stalked, equal in size, and usually but not always divided into two rounded lobes, 5–18 mm long. Flowers are in clusters of one to five. The lower halves of the stalks of ripening fruits bend back, so that there is a sharp angle about half-way along the stalks.

It lives on roadsides, and in waste and cultivated areas: flowers all the year round.

Another common yellow-flowered oxalis is Bermuda buttercup (*O. pes-caprae*★); this has no above-ground stems, but has underground stems producing leaves and numerous bulbils at its apex. The leaflets are unstalked, the terminal leaflet being larger than the other two. The flowers are larger (25–30 mm).

*Oxalis incarnata**

Plate 37

LILAC OXALIS

15–25

A low to short perennial, with fleshy, contractile root and bulb; above-ground stems produce bulbils. Leaflets are more or less unstalked, divided by a broad cut into two rounded lobes, and have a row of orange calli beneath. The flowers are solitary, the sepals having two calli at their tips.

Poppy Family – Papaveraceae

H O S – R
K 2 see below
C 4
M many
F (2–many)

The Californian poppy, common in New Zealand and described below, is an exception, since its sepals are united and its ovary is not completely superior. In most genera of the family the flowers are large, with brightly coloured petals, but they do not usually produce nectar: insects visit the flowers to collect pollen.

Many of this family produce poisonous latex. The opium poppy is the only economically important member of this family, as the drug opium is obtained from its latex. The seeds of the plant are used in baking, but are harmless since they contain no opium. Poppy-seed oil is used in foodstuffs and in paint and soap making. Many species are grown as ornamental garden plants, including Californian poppy and *Meconopsis*.

*Eschscholzia californica**

Plate 37

CALIFORNIAN POPPY

30–50

A low to medium, hairless annual or perennial with feathery leaves. The sepals are fused to form a pointed hood, which drops off when the flower-bud opens. The anthers are much longer than the filaments.

It lives in dry areas beside roads and on waste land: flowers from October to February.

Passion Flower Family – Passifloraceae

LHS A C + R
K 3–5 fused at the base
C 3–5
M 3–5 sometimes up to 10
F <u>(3–5)</u>

These are mostly climbing plants, with tendrils in the axils of their leaves. The flowers are more complicated in structure than usual. The sepals and petals are often similar in appearance, so that in *Passiflora*, for example, there appear to be 10 petals (the 'Ten Commandments'). The stamens are usually mounted on a central stalk arising from the base of the flower. Also on this stalk are numerous filaments, forming a radiating corona (the 'crown of thorns'), and on the end of the stalk is the ovary. There is usually one style, which may branch.

Passionfruit are widely grown in tropical areas for the juicy fruit, which is also marketed as canned fruit juice and as a flavouring for ice cream. The showy flowers make passion flower an attractive garden plant in warmer areas or in sunny spots in gardens in temperate regions.

*Passiflora edulis** — Plate 19

BLACK PASSIONFRUIT — **40–50**

A climbing, perennial plant with tendrils. Leaves are three-lobed, dark green and shiny above, and the style branches into three (the 'three nails'). The fruit is purple when ripe.

It grows in scrub and forest margins close to passionfruit plantations or former plantations in north Auckland, and may escape and become established in other areas: flowers from July to March.

*Passiflora mollissima** — Plate 19

BANANA PASSIONFRUIT — **20–40 long**

A perennial climber to several metres. Leaves are three-lobed, and the stipules are 5–10 mm wide, wider than the stem (see photograph, top right). Showy, pink flowers hang downward. Fruits are green at first, then ripen to dull yellow, looking like small, straight bananas.

It climbs on roadside shrubs, and flowers throughout the year.

The northern banana passionfruit (*P. mixta**) is common in north Auckland and a few other areas. It has larger flowers, and its stipules are only 2–4 mm wide, about as wide as the stem. The fruit of both plants is edible.

Inkweed Family – Phytolaccaceae

HSfewT A E – R
P (4–10) calyx
M 4–10 alternate with tepals
F 2–16 free or slightly joined

A small family mainly from tropical and sub-tropical America.

*Phytolacca octandra** — Plate 16

INKWEED — **5–7**

A medium to tall, perennial plant, becoming woody when mature. The leaves are elliptical, entire, with a pointed tip; stems are often reddish; flowers are inconspicuous, whitish and quickly fruiting. The inflorescence is a tall, narrow raceme of rounded, shiny green fruits, becoming purplish black on ripening.

It lives on waste and cleared ground: flowers from November to August.

The first part of the genus name comes from Greek *phyton* (plant), and the second part comes from Latin *lacca* referring to the lac insect; this insect is one from which a dye is obtained. A red dye is obtained from this plant too, so its name describes it as a plant version of the lac insect.

Plantain Family – Plantaginaceae

H AOB Si – R
K (4)
C (4) membranous
M 4 see below
F (2)

Members of this family all have a long, leafless stem bearing an inflorescence, and arising from a basal rosette of leaves. The leaves have parallel veins, narrowing gradually into the leaf stalk, which is slightly winged. The inflorescence is usually a spike of small, inconspicuous flowers. The stamens have long filaments and project conspicuously from the inflorescence; they produce large amounts of dry, dusty pollen.

There are no members that are economically important in the positive sense, but the family includes many members that are persistent garden weeds.

Plantago lanceolata★ **Plate 17**

NARROW-LEAVED PLANTAIN – RIBWORT **4**

A short to tall, perennial plant with a basal rosette of lance-shaped leaves, with three to seven prominent veins. The spike is blackish and oval, being about 20 mm (can be up to 60 mm long), at the top of a grooved stalk. Anthers are creamish white when they first open, later turning brown.

It lives in a wide variety of open, disturbed areas: flowers from July to April.

Plantago major★ **Plate 17**

BROAD-LEAVED PLANTAIN **4**

A low to tall, perennial plant with a basal rosette of broad leaves, with five to seven prominent veins. The leaf stalk may be almost as long as the leaf blade. The spike is greenish and elongated, being 15–300 mm long – as long as or longer than its stalk. Anthers are pale purple when they first open, later turning brown.

It lives in a wide variety of open, disturbed areas, particularly damp areas.

This is named after the Latin *planta*, meaning 'foot', because of the way the leaves lie flat to the ground.

Phlox Family – Polemoniaceae

H AO SiC – R
K (5)
C (5) bell-shaped, funnel-shaped or saucer-shaped
M 5 alternate
F (3)

The only well-known members of this family are garden plants such as *Phlox*, *Polemonium* and *Cobaea*.

Navarretia squarrosa★ — Plate 31
CALIFORNIAN STINKWEED — **5–8**

A hairy, sticky, unpleasant-smelling annual of medium height. The feathery leaves soon fall off, leaving branches bearing compact inflorescences. The bracts of the inflorescence and the sepals end in spines.

It is found on waste land, beside roads, and in other open habitats: flowers from November to April.

Dock Family – Polygonaceae

HfewST A Si + R
K see below
C see below
M 6–9
F 1

The flowers have two types of arrangement of the sepals and petals:

- in two separate whorls – three sepals and three petals
- in a spiral – two sepals and one segment formed by fusing a sepal and a petal plus two petals

The petals and sepals are similar in appearance. A distinctive feature of this family is the sheath surrounding the stem at the base of each leaf-stalk; this ochrea is formed by the fusion of stipules. It can be membranous or green, with or without a fringe of narrow teeth, and variously shaped. The ochrea is often important in identification, as the flowers of many genera are small and very similar. The stems are often joined and swollen where the leaves are attached.

Food-plants in this family include:

- Rhubarb: the leaf stalks are cooked and eaten; they can also be made into wine.
- Buckwheat: starchy seeds.
- Sorrel: the leaves are cooked and eaten as a vegetable.

Several species of *Polygonum* are grown as ornamental plants, including the vigorous climber Russian vine (*P. baldschuanicum**). The genus *Polygonum* also includes a number of common weeds of gardens and agricultural land, as does the genus *Rumex* (docks and sorrels).

Polygonum aviculare agg.* — Plate 16

MAKAKAKA – WIREWEED – KNOTGRASS — 2–3

A low, hairless, mat-forming annual, with wiry, 'jointed' stems widely spreading close to the ground. It has lance-shaped leaves, those on the young and main stem being much larger than those on the branches. There are clusters of one to six flowers in the axils of the leaves, and the flowers are white or green with reddish borders. The ochreae are silvery, with jagged edges and few veins.

It is very common on waste and cultivated ground, from sea-level to 700 m: flowers from November to June.

In China and Japan this plant was used as a source of a blue dye, similar to indigo. The stems contain large quantities of silica crystals, making the plant a useful scouring-pad for dishes. A more bizarre use of the plant in the Middle Ages was to stunt the growth of children so that they could become dwarfs in circuses. This was known to William Shakespeare who, in *A Midsummer Night's Dream*, wrote:

> Get you gone, you dwarf,
> You minimus of hindering knotgrass made;
> You bead! You acorn!

The abbreviation 'agg.' is for aggregate, indicating that there is a confusing assortment of different forms of this plant, which are difficult to separate out into distinct subspecies or varieties.

In small-leaved wireweed (*P. arenastrum**) the leaves on the young and main stems are about the same length as, or only slightly longer than, the leaves of the branches.

WILLOWHERBS, FLAXES, LOOSESTRIFES

Epilobium billardiereanum
p. 135

Epilobium glabellum
p. 135

Epilobium tetragonum
p. 136

Linum bienne
p. 130

Linum trigynum
p. 130

Lythrum hyssopifolia
p. 132

Lythrum junceum
p. 132

PLATE 34

MUSTARD FAMILY

Brassica rapa
p. 88

Cakile maritima
p. 88

Capsella bursa-pastoris
p. 89

Cardamine debilis
p. 89

Hesperis matronalis
p. 90

Cardamine hirsuta
p. 89

Coronopus didymus
p. 89

Raphanus raphanistrum subsp. *maritimus* p. 91

Lunaria annua p. 90

Rorippa sylvestris p. 92

Notothlaspi australe p. 91

Rorippa nasturtium-aquaticum p. 92

Diplotaxis muralis p. 90

Sisymbrium officinale p. 92

PLATE 36

Buttercups

Ranunculus acris
p. 151

Ranunculus enysii
p. 152

Ranunculus bulbosus
p. 152

Ranunculus sceleratus
p. 154

Ranunculus lyallii
p. 154

Ranunculus insignis
p. 153

Ranunculus parviflorus
p. 154

Buttercups, sorrels, poppy

Ranunculus flammula
p. 152

Ranunculus sericophyllus
p. 155

Ranunculus repens
p. 154

Oxalis articulata
p. 138

Oxalis corniculata
p. 138

Oxalis incarnata
p. 139

Eschscholzia californica
p. 139

PLATE 38

Cushion plants

Phyllachne colensoi
p. 170

Neopaxia australasica
p. 148

Hectorella caespitosa
p. 125

Raoulia australis
p. 69

PETER JOHNSON

Raoulia grandiflora
p. 69

Raoulia subsericea
p. 70

*Polygonum hydropiper**
Plate 16

WATER PEPPER
2.5–3

A medium to tall, erect, hairless annual. Its leaves taste strongly peppery and have small, translucent spots on them; the leaves are 10–120 mm long and 3–20 mm wide. Ochreae are up to 20 mm long, reddish brown, possibly with a fringe of a few bristles. The inflorescence is a narrow raceme, which partly curls downward, with flowers well spaced out. Flower stalks are short, hidden inside the bracts until the fruits are maturing.

It grows in damp areas, beside streams, ponds and lakes, and on poorly drained land: flowers from November to June.

Hybrids occur between this species and willow weed. A plant of similar appearance is *P. punctatum**; its leaves are larger (40–150 mm long, 10–40 mm wide) and have a slight peppery taste. Its ochreae have a fringe of bristles. The flowers are white or pale green, with flower-stalks usually longer than the bracts.

*Polygonum persicaria**
Plate 16

WILLOW WEED
2–3

A medium, spreading annual with lance-shaped leaves, usually with black areas and spots on the upper surface. The inflorescence is an erect, dense spike of pink flowers, less than 40 mm long. Ochreae are green or pink, with a fringe of hairs.

It grows on waste and cultivated land near to habitation: flowers year round.

Hybrids occur between this species and water pepper. According to folklore, the dark spots on the leaves show where the Devil pinched them, because they lack the hot peppery taste of the leaves of water pepper. Shetlanders use this plant as a source of a yellow dye.

*Polygonum polystachyum**
Plate 16

INDIAN KNOTWEED
2.5–4

A tall, densely growing perennial with stout stems and leaves up to 90 mm long. The main vein is red and the side veins of the leaves are pink beneath. The inflorescences have thin red branches.

It lives on roadsides and waste areas near habitation: flowers from November to May.

The name *Polygonum* means 'many knees', referring to the jointed appearance of the stems.

Polygonum strigosum★ **Plate 16**

POLYGONUM **2–2.5**

A scrambling, bristly-hairy annual, with arrow-shaped leaves, easily distinguished from the other members of this genus.

It is found in damp areas in the northern parts of the North Island: flowers from January to February.

Rumex acetosella★ **Plate 17**

SHEEP'S SORREL **2**

A low to short, hairless, reddish perennial with leaves arrow-shaped or very narrow, with two spreading or forward-pointing lobes at the base. The leaves have a bitter taste. The inflorescence is a narrow raceme of reddish or yellowish flowers.

It is abundant on open waste land from sea-level to 1500 m, particularly in poor, dry habitats: flowers throughout the year.

The name 'sorrel' is derived from the French word *surele*, meaning 'sour'.

Rumex conglomeratus★ **Plate 17**

CLUSTERED DOCK **2**

A medium to tall perennial with broad, stalked leaves (not as large as those of curled dock), with finely wavy edges. The inflorescence consists of whorls of greenish flowers, the whorls being spaced apart from each other.

It lives in damp areas such as pastures, and beside lakes, streams and ditches, and swamps, especially where rainfall is high: flowers from November to April.

Rumex crispus★ **Plate 17**

CURLED DOCK **2**

A medium to tall perennial with broad, stalked leaves with finely wavy edges. The inflorescence consists of whorls of green flowers, the whorls being very close together, almost touching.

It lives in damp areas such as pastures, beside lakes, streams and ditches, and in swamps, and occasionally on drier sites too: flowers from November to April.

Rumex sagittatus★ **Plate 17**

CLIMBING DOCK **2**

A climbing or scrambling perennial with stems up to 3 m long. The leaves are triangular or spear-shaped. The photograph is taken at the fruiting stage.

It lives in coastal areas, on waste land and banks, often climbing on hedges, and in gardens: flowers from November to March.

Purslane Family — Portulacaceae

H AO E - R
K 2 or (2) lower sepal overlaps upper sepal
C 5 may be united at the base
M 3–many, usually five, opposite the petals, or two whorls of five
F (3)

The leaves are often fleshy, and petals often satiny.

Species of a few genera, including *Portulaca* and *Calandrinia*, are grown as ornamental plants.

Calandrinia compressa★ **Plate 21**

CURNOW'S CURSE **12–15**

A short to medium annual, with a basal rosette of narrowly oval leaves sheathing the stem. The stem leaves and bracts are similar, but lacking stalks. The two sepals are triangular, large (about 7 mm × 7 mm), and joined together at the base. Stigmas are purple.

It is found on waste and cultivated land: flowers from October to March.

Calandrinia menziesii★ **Plate 21**

CURNOW'S CURSE **18–20**

A short to medium annual, with a basal rosette of oval to spoon-shaped leaves, sheathing the stem. Stem leaves and bracts are similar, with stalks. The two sepals are spear-shaped, large (about 7 mm × 5 mm) and free at the base. Stigmas are white.

It is found on waste and cultivated land: flowers from October to December.

Neopaxia australasica — Plate 38

NEOPAXIA — 20

A low, creeping and rooting, mat-forming, fleshy perennial, with flowering stalks up to 30 mm long. The narrow leaves are fleshy and from 10 to 50 mm long. The leaf-stalk is expanded into a papery sheath, which clasps the stem. The stamens are opposite the petals, helping to distinguish this from similar white-flowered plants.

It is a widespread plant, living beside streams and in other damp habitats, from lowlands to 2000 m. It flowers from November to January.

This genus has also been known as *Claytonia* and as *Montia*.

*Portulaca oleracea** — Plate 16

PURSLANE — 6

A low to short, spreading annual, often with a pinkish tinge. The leaves are fleshy, oval and often tinged pink. The small, yellow flowers are in ones, twos or threes in the leaf axils.

It grows on waste and cultivated land: flowers from November to March.

The plant is rich in vitamin C and was at one time used to prevent scurvy.

Primrose Family – Primulaceae

H OWB E – R
K (5)
C (5)
M 5 opposite
F <u>(5)</u> or half-inferior

These plants often have a rhizome or tuber, from which arises a tight cluster of leaves.

Flowers are solitary or clustered at the end of long, leafless stalks.

The genus *Primula* includes many plants grown for ornament. Other ornamental genera in this family are *Cyclamen* and *Lysimachia*.

*Anagallis arvensis** — Plate 21

SCARLET PIMPERNEL — **8**

A low-growing annual with square stems. The leaves have black dots on the underside. The brick-red flowers, with the sepals clearly visible between the petals, distinguish this species from all others.

It is very common and widespread on open waste and cultivated land: flowers all the year round.

This is sometimes known as the 'poor man's weather-glass', as the flowers are said to close just before it rains. The plant shown in the photograph belongs to the common subspecies *arvensis*, and to the red-flowered variety *arvensis*, so that its full name is *Anagallis arvensis* subsp. *arvensis* var. *arvensis*.

Anagallis arvensis subsp. *arvensis* var. *coerulea** — Plate 21

BLUE PIMPERNEL — **8**

The plant is the same as scarlet pimpernel, except that the petals are blue, with edges more toothed. The photograph shows both varieties.

This is less common than the scarlet variety, and rare south of Nelson.

Samolus repens — Plate 32

MAAKOAKO — **6–7**

A spreading and rooting, hairless perennial, with stems up to 400 mm long. The leaves are narrow to spoon-shaped, and fleshy. The ovary is half-inferior.

It grows on salt marshes and rocky coastal areas: flowers from November to February.

Buttercup Family – Ranunculaceae

HL AO SiC - R
K various, see below
C various, see below
M many and indefinite
F many and indefinite

Although most of the family are herbs, it also includes some woody climbers.

The parts of the flower are all free and spirally arranged. Numbers of parts vary over a wide range and are not necessarily constant for any given species. There may be distinct sepals and petals, but in some species the sepals are petal-like.

Several genera provide us with ornamental plants, including *Ranunculus*, *Clematis*, *Aquilegia* (columbine), *Helleborus* (hellebore) and *Delphinium*. Members of this family often contain alkaloids, making them poisonous.

Clematis paniculata — Plate 19

PUAWHANANGA – PIKIARERO — 50–60

A tall, woody, climbing perennial with opposite leaves (not alternate, as in most members of this family) of three leaflets. The flowers on a plant are either all male or all female; they are in clusters of six or more, each usually with six white, petal-like sepals. The number of sepals may vary between five and eight. There are no petals. Male flowers have over 50 stamens, with mauve anthers. Female flowers are slightly smaller, with a few infertile stamens and numerous carpels.

It lives in lowland and low montane forest, more commonly around the margins: flowers from August to November.

*Clematis vitalba** — Plate 19

OLD MAN'S BEARD – TRAVELLER'S JOY — 16–20

A tall, woody, climbing perennial with opposite pinnate leaves, usually with five leaflets; leaf and leaflet stalks twine around supports. Flowers occur in the axils of the leaf and at the end of the stem. The flowers have four or five petal-like sepals and no petals. The sepals are white inside and greenish cream with a white border outside. The flowers have a pleasant smell. Flowers have both male and female parts. The fruits of this species give it one of its names, as each fruit has a long, persistent style densely covered with long, greyish hairs.

It lives in scrub, hedges and the margins of forests: flowers from March to October.

Many people call this plant by its scientific name – clematis – instead of using its common names. But there are frequent arguments as to how the name should be pronounced: whether it is 'clematis' (short 'e'), 'cleematis' (long 'e') or 'claymatis' is a matter of opinion. The first is suitable for ordinary gardeners' chat, though either one of the other pronunciations is more correct for scientific use. Whatever version is used, the accent should be on the first syllable, not on the second.

IDENTIFYING INTRODUCED BUTTERCUPS

1 Leaves not lobed or divided ▸ ***Ranunculus flammula*** (p. 152)
  Leaves lobed or divided ▸ 2

2 Leaves divided into three stalked leaflets, the stalk of the middle leaflet being longer than that of the other two leaflets ▸ 3
  Leaves not divided as above ▸ 4

3 Plant with creeping, rooting stems, sepals spreading ▸ ***R. repens*** (p. 154)
  Plant with bulb at base of stem, sepals turned back ▸ ***R. bulbosus*** (p. 152)

4 Sepals turned back ▸ ***R. sceleratus*** (p. 154)
  Sepals spreading ▸ 5

5 Flowers less than 8 mm in diameter ▸ ***R. parviflorus*** (p. 154)
  Flowers more than 8 mm in diameter ▸ ***R. acris*** (below)

A fairly common plant that looks like a buttercup is Indian strawberry (p. 158), which belongs to the Rose Family.

*Ranunculus acris** — Plate 36

GREAT BUTTERCUP — **15–25**

A medium to tall perennial with rounded leaves, deeply three- to seven-lobed, the lobes being pinnately cut into narrow segments with pointed ends; the end lobe has no distinct stalk. The flower-stalk is rounded in section. The spreading sepals are yellow, with short, closely pressed hairs.

It is found in damp paddocks, gardens, lawns, verges, waste ground and on the edges of swamps: flowers from December to April.

This and many other buttercups are poisonous to stock.

Ranunculus bulbosus★ Plate 36

BULBOUS BUTTERCUP 15–25

A short perennial growing from a swollen stem base. The stems are hairy, the hairs being pressed to the stem. The leaves are divided into three deeply toothed lobes, and the end lobe is stalked. The sepals are bent backward so that their tips are pressed closely to the flower-stalk.

It is found in damp areas, pasture, waste ground and roadsides: flowers from October to December.

A similar plant is hairy buttercup (*R. sardous*★), but this lacks the bulbous stem base. It is more hairy than the bulbous buttercup, and the hairs spread out from the stem. Its flowers are pale yellow and its fruits are flattened, with small, blunt swellings near the edges.

Ranunculus enysii Plate 36

BUTTERCUP 15–30

A short to medium perennial growing from a rhizome. The leaves vary in shape. Usually they are oval, divided into three to five rounded, shallow-toothed or three-lobed leaflets. The leaflets are stalked and often have red veins and margins. In some areas, the leaves are rounded and three-lobed. The flower-stalks usually have only one flower, rarely up to three.

It is found on the ranges of the South Island, living in damp, sheltered locations among tussocks or in rock-clefts, at 900 m to 1500 m. Flowers in October and November.

Ranunculus flammula★ Plate 37

SPEARWORT 6–20

A short to medium, creeping perennial, with rather thick, fleshy stems, usually reddish tinged. The leaves are simple, lance-shaped and shallowly toothed. The flower-stalk is grooved.

It lives in wet locations: flowers from October to March.

The name *flammula*, meaning 'little flame', refers to the burning taste of its poisonous leaves.

The buttercups

Buttercups are all members of the genus *Ranunculus*, named from a Latin word meaning 'little frog': this refers to the occurrence of this genus in predominantly damp habitats. Pastures, swamps and streamsides are the places where buttercups are most likely to be found. Some members of the genus (often called crowfoots) are fully aquatic, rooting in the beds of streams and having leaves specially adapted to a submerged existence.

The 'typical' buttercup has five free green sepals, five free yellow petals, a large number of stamens, spirally arranged, and a large number of carpels. There are some exceptions to these rules, for example, the number of petals varies, and petals may be white, as in two of the New Zealand native species. There is usually a small nectary scale inside each petal, at or near its base. Leaves are usually stalked and tend to be rounded in outline. They are often divided palmately into lobes or segments.

There are 47 species of buttercup growing wild in New Zealand, falling into three main groups. The ones you are most likely to find are the introduced species, of which there are 13; these are weeds of cultivation and occur widely in pastures and crops and on roadsides and waste land. The 34 native species are mostly restricted to mountain areas and offshore islands. There are also a few native species that grow in lowland areas. On p. 151 we give a key to the introduced species.

Ranunculus insignis — Plate 36

KORIKORI — **20–30**

A short, tufted perennial with oval to circular leaves, which are thick and hairy, usually on both surfaces. The leaf margin has rounded teeth and is fringed with brown hairs. There are 2–10 flowers on a stem, and the flowers have five to seven petals.

It grows in tussock grassland, scrub and rocky outcrops, from 700 m to 1800 m: flowers from November to February.

Ranunculus lyallii **Plate 36**

MOUNT COOK LILY – GREAT MOUNTAIN BUTTERCUP 50–80

A short to tall, tufted perennial with circular, peltate leaves up to 400 mm in diameter. The leaves are hairless on both surfaces and the margin has rounded teeth. There are 5–15 flowers on a stem, and they have 10–16 petals.

It is found only in the South Island, growing beside streams and in other wet situations from 700 m to 1500 m. It flowers in December and January.

This is the largest buttercup in the world. The common name 'lily' is botanically incorrect: it does not even belong to the same class!

*Ranunculus parviflorus** **Plate 36**

SMALL-FLOWERED BUTTERCUP 3–6

A short, annual, hairy plant with rounded leaves partly cut into three to five deeply toothed lobes. The flower-stalk is grooved, with several flowers on each stalk. The sepals are bent backward, but their tips do not touch the flower-stalk; petals and sepals are about the same length. The fruits are covered, with hooked spines.

It lives on waste ground, in tussock grassland and scrub, and in gardens: flowers from October to December.

*Ranunculus repens** **Plate 37**

CREEPING BUTTERCUP 20–30

A short to medium perennial with runners, which arch over and take root at intervals. The lower leaves are divided into three stalked segments, each of which is divided again into three toothed lobes. Sepals are spreading.

It lives in wet situations beside roads, in ditches, on river banks and in other waste areas: flowers from November to January.

*Ranunculus sceleratus** **Plate 36**

CELERY-LEAVED BUTTERCUP 5–15

This short to medium annual is hairless or sparsely covered with fine, closely pressed hairs on its stems, leaves and flower-stalks. Its leaves are oval to kidney-shaped and look like celery leaves, being deeply cut into narrow lobes. There are up to 30 flowers on each stem. The boat-shaped sepals are turned slightly back. The fruiting head is cylindrical, as seen in the photograph, and bears several hundred fruits.

It grows in damp mud around ponds, on roadsides and on waste land: flowers from October to February.

Ranunculus sericophyllus — Plate 37

YELLOW MOUNTAIN BUTTERCUP — 30–50

A low to short, tufted perennial with oval leaves in three segments, the segments again being deeply cut into three, giving the leaves a feathery appearance. The leaves are clad in silky hairs, as are the broad sepals. The flower-stalks bear a solitary flower, with five to eight petals.

It is found in the South Island, in wet, rocky locations in fellfields, from 1400 m to 2100 m.

Mignonette Family – Resedaceae

HS A Si - I
K 4–8 may be joined near the base
C 4–8
M 3–40
F (2–7)

Weld (see below) was formerly important for its yellow dye. Common mignonette is sometimes grown in gardens for its pleasant perfume; the leaves retain their perfume when dried and, for this reason, the Ancient Egyptians placed dried mignonette plants in tombs beside the mummies. When he was fighting in Egypt, Napoleon discovered the plant and sent seeds to Josephine in France, where the plant became popular and soon spread over Europe.

*Reseda luteola** — Plate 44

WILD MIGNONETTE – WELD – DYER'S ROCKET — 4–5

A medium to tall biennial with narrowly lance-shaped, entire, wavy-edged leaves and a single stem bearing a raceme of green-yellow flowers. The flowers have four sepals and petals.

It grows in waste areas, beside roads and in gardens: flowers from November to February.

Two other mignonettes frequently occur, in which some of the leaves are pinnately lobed. White mignonette (*R. alba*★) has white flowers with five to six sepals and petals. Cut-leaved mignonette (*R. lutea*★) has yellow flowers with six sepals and petals.

Rose Family — Rosaceae

TSH AB SiC + R
K 4–5 often with epicalyx (see Indian strawberry, for example)
C 4–5
M many in whorls of five, or large indefinite number
F 1–many or 1–few sometimes half-inferior

A wide variety of plants comprise this family, ranging from low, creeping herbs to moderate-sized trees. Often the stipules are fused to the lower end of the leaf-stalk. The family has a wide range of fruits, from fleshy drupes with a woody stone (for example, plum) to a 'false fruit' consisting of many small, dry achenes scattered on a swollen, fleshy receptacle (for example, strawberry). In some, the achenes may be enclosed in a swollen, fleshy receptacle, forming a pome (for example, apple).

The edible fruits produced by the Rose Family include apple, pear, quince, medlar, cherry, plum, apricot, peach, nectarine, almond, blackberry, raspberry, loganberry, boysenberry, strawberry and loquat. Varieties of cherry, plum and almond are also grown for their prolific spring blossoms. Hundreds of varieties of the genus *Rosa* have been bred with a wide variety of form, colour and fragrance. Other popular flowering shrubs of this family include *Cotoneaster*, *Chaenomeles*, *Spiraea*, *Sorbus* and *Pyracantha*. Ornamental herbs include various geums and potentillas.

Acaena agnipila★ — Plate 39

BIDIBIDI — **3–5**

A short to tall, erect perennial, with pinnate leaves, 80–150 mm long. There are 8–13 pairs of toothed leaflets. The inflorescence is a spike. The flowers have five green sepals but no petals. The mature calyx tube (see note below) has 12–55 red spines, 1–2 mm long, with barbed tips (see photograph of *A. inermis* on plate 39). *A. agnipila* differs from other members of the genus in being erect rather than prostrate, and having a narrow spike instead of a rounded inflorescence.

It lives in dry grassland and waste areas: flowers from October to March.

In *Acaena* the bases of the sepals unite to form a calyx tube, partly surrounding the carpels. When the carpels mature, prominent spines, often with barbed tips, develop on the outside of the calyx tube. These spines aid dispersal of the fruits in the coats of furry animals. The name *Acaena* comes from Greek *akaina*, referring to the thorny fruits. There are 14 species of *Acaena* native to New Zealand, and two introduced species. Many of the natives have rather restricted distribution.

Acaena anserinifolia — Plate 39

PIRIPIRI – BIDIBIDI — **1**

A prostrate, hairy perennial with pinnate leaves 10–75 mm long. There are four to six pairs of toothed leaflets; the end leaflet is longer than it is wide. The inflorescence is a globular cluster of 50–60 flowers, which have four green sepals but no petals. The two stamens with white anthers and the white style give the inflorescence an overall whitish appearance. The mature calyx tubes develop four pale brown spines 4–9 mm long, with barbed tips, radiating from the inflorescence, as in the photograph of *A. inermis* on plate 39.

It is found at forest margins and in shrubland from lowland to subalpine regions: flowers from December to April.

The leaves may be used as tea. This species is unusual in that it has managed to become naturalised in Britain, by way of Australia, carried as fruits in raw wool.

Acaena inermis — Plate 39

BIDIBIDI — **1.5–4**

A prostrate, hairless, mat-forming perennial, with pinnate leaves 15–70 mm long. There are five to seven pairs of toothed leaflets. The leaves are dull, with colour varying from greyish blue-green, to purplish brown (as in the photograph) to light olive-green. The inflorescence is a globular cluster of about 20 flowers. The flowers have four light green sepals, but no petals. The mature calyx tube has one to four red spines up to 13 mm long, without barbs.

It lives in open ground, grassland and fellfield up to 1600 m, and flowers from

November to January. The red spines give the mat a distinctive appearance from January to April.

The name *inermis* means 'no thorns'.

*Duchesnea indica** — Plate 39

INDIAN STRAWBERRY — 10–12

A low to short, mat-forming perennial with strawberry-like trefoil leaves. The solitary flowers look at first glance like buttercups, but the prominent epicalyx, with five leafy toothed segments, confirms its identity.

It lives in shady, damp places on forest margins and in light shade in reserves and waste land: flowers from July to April.

*Fragaria vesca** — Plate 39

ALPINE STRAWBERRY — 12–18

A low to short perennial with runners. The leaves are trefoil, bright green, with long, silky hairs on the upper surface. The flowering stems have 3–10 long-stalked flowers. The flower has five triangular sepals, surrounded by an epicalyx of five narrow entire segments. The fruit is a smaller version of the commercial strawberry.

It lives in damp areas, on roadsides, waste land, and in clearings in forests, from sea-level to 1000 m: flowers from November to April.

*Potentilla recta** — Plate 39

TALL CINQUEFOIL — 20–25

A short to medium, shaggily hairy perennial with stiff, erect stems. The leaves are rounded and palmate; the lower leaves have five to seven deeply toothed leaflets. Epicalyx segments are narrowly lance-shaped, and petals are longer than the sepals, and deeply cut. The pale (sulphur) yellow of the petals distinguishes this from other cinquefoils.

It is found in the South Island, growing in dry locations in grassland, on roadsides and in waste areas: flowers from December to May.

Rubus australis Plate 40
BUSH LAWYER 8–15

A scrambling shrub with thorny stems. Leaves are palmate, usually with three or five leaflets, with small, reddish thorns. The terminal leaflet is oval to rounded, with the leaflet stalk longer than the leaflet, and with prickly teeth on each side. The flowers have either stamens or carpels, but not both. The photograph is of a male plant, which has only stamens in its flowers. Fruits are yellowish to orange.

It is found in lowland to montane forest and scrub: flowers from August to January.

This plant and the one below are well known to New Zealanders because of their painful thorns.

Rubus cissoides Plate 40
BUSH LAWYER 8–15

A scrambling shrub with thorny stems. The leaves are palmate, usually with three or five leaflets, with stout, reddish thorns. The terminal leaflet is lance-shaped, with the leaflet stalk usually much shorter than the leaf, and with prickly teeth on each side. The flowers have either stamens or carpels, but not both. The photograph is of a male plant, which has only stamens in its flowers. Fruits are orange to red.

It grows in lowland to montane forest, scrub and open areas: flowers from August to December.

Rubus fruticosus agg.* Plate 40
BLACKBERRY 20–30

A tall, scrambling shrub with thorny stems. The tips of the stems arch over, rooting where they touch the soil. The leaves are usually trefoil, but some have five leaflets; the plant is thorny. Flowers are pink or white. Fruits, consisting of several rounded, fleshy segments, are red at first, ripening to black.

It grows in a wide variety of habitats, from sea-level to 1000 m: flowers from November to May.

Bedstraw Family – Rubiaceae

STfewH O E + R
K 4–5 often reduced, e.g. to a ring of tissue
C (4–5)
M 4–5 alternate
F (2)

The leaves are arranged so that a pair of leaves is at right angles to the pairs above and below it. In some genera the stipules are large and leafy, so that it appears that the plant has whorls of leaves.

Plants of economic importance in this family produce coffee, quinine and the medicament ipecacuanha. Among the ornamental garden plants are gardenia and coprosma.

*Galium aparine** **Plate 29**

CLEAVERS – GOOSE GRASS **1–1.5**

A medium to tall, climbing annual with square stems, the angles having stiff, hooked hairs. There are stiff hairs on the leaf margins and on the underside of the midrib of the leaves, so that the whole plant tends to cling to clothing or to the fur of animals. The leaves are apparently in whorls of six to eight, with pointed tips. There are minute flowers in few-flowered clusters, the stalks of the clusters being longer than the leaves. The fruits also bear hooked hairs, which help in their dispersal.

It grows where it can obtain support from other plants, in waste areas, forest margins, weedy gardens and pastures: flowers from July to March.

The plant is called 'goose grass' because it was at one time fed to newly hatched goslings. The name 'cleavers' refers to its tendency to cling to passers-by. Two other names, 'hug-me-close' and 'sweethearts' also refer to this clinging habit.

*Galium palustre** **Plate 29**

MARSH BEDSTRAW **1.5–2**

A short to medium perennial without hooked hairs, though the angles of the stem are rough to the touch. The leaves are apparently in whorls of four to five and blunt-tipped; leaves and stipules become black or dark brown when they dry. Flowers are

Rose Family

Duchesnea indica
p. 158

Acaena agnipila
p. 156

Acaena anserinifolia
p. 157

Acaena inermis
p. 157

Fragaria vesca
p. 158

Potentilla recta
p. 158

PLATE 40

Nightshades, brambles

Datura stramonium
p. 168

Rubus australis
p. 159

Rubus fruticosus agg.
p. 159

Rubus cissoides
p. 159

Solanum laciniatum
p. 169

Solanum nigrum
p. 169

Orobanche minor
p. 137

Digitalis purpurea
p. 162

Antirrhinum orontium
p. 162

Scrophularia auriculata
p. 166

Euphrasia cuneata
p. 162

Linaria arvensis
p. 163

Linaria repens
p. 163

Euphrasia zelandica
p. 162

PLATE 42

Foxglove Family

Parahebe catarractae
p. 165

Parahebe lyallii
p. 165

Ourisia macrocarpa
p. 164

Ourisia sessilifolia subsp. *splendida*
p. 165

Hebe lycopodioides
p. 163

Parentucellia viscosa
p. 166

FOXGLOVE FAMILY, EVENING PRIMROSES

Veronica persica
p. 167

Verbascum thapsus
p. 166

Verbascum virgatum
p. 167

Veronica arvensis
p. 167

Oenothera stricta
p. 136

Veronica serpyllifolia
p. 167

Oenothera glazioviana
p. 136

PLATE 44

NASTURTIUM, VALERIAN, VERBENAS, MIGNONETTE

Tropaeolum majus
p. 171

Centranthus ruber
p. 173

Lantana camara
p. 174

Verbena litoralis
p. 174

Reseda luteola
p. 155

Verbena bonariensis
p. 174

in clusters of 10 or more, and the branches of the inflorescence spread apart widely as the fruits mature.

It grows in wet, shady places: flowers from November to April.

Slender marsh bedstraw (*G. debile**) has a less open inflorescence, its branches not spreading as widely, so that the stalks are more erect. Its leaves and stipules remain green or turn light brown on drying.

*Sherardia arvensis** — Plate 29

FIELD MADDER — **2–4.5**

A low to short, spreading annual with square stems, its lower leaves elliptical and apparently in whorls of four, and upper leaves lance-shaped in whorls of five to six. The flowers are in few-flowered, rounded heads surrounded by 8–10 leafy bracts.

It lives in grassy places from sea-level to 500 m: flowers all the year round.

Foxglove Family – Scrophulariaceae

HSfewT AOW Si – I
K (5) (4) in *Veronica*, sometimes two-lipped
C (5) (4) in *Veronica*, often two-lipped
M 5 alternate, sometimes two or four
F (2)

This widespread and commonly occurring family consists mostly of herbs and low-growing shrubs (such as *Hebe*). The genus *Hebe* is an extremely large one, containing about 100 species, most of which are indigenous to New Zealand. Since they are shrubs they are outside the scope of this book, but we include one whipcord hebe, as a representative of this important genus, since this type of hebe may at first glance be taken for a herbaceous plant.

The foxglove is the only family member of economic importance, since it is the source of the alkaloid digitalin, used to treat heart disease. The foxglove is also grown as an ornamental plant, as are many species of hebe, and herbs such as snapdragon, *Mimulus*, *Nemesia*, speedwell and mullein.

Antirrhinum orontium★ **Plate 41**

LESSER SNAPDRAGON **12–18 long**

A short to medium annual with an upright, usually unbranched stem and narrow, entire leaves. The calyx is 10–18 mm long, divided almost to the base into narrow segments of unequal length.

It lives in waste areas and on roadsides: flowers from July to March.

The garden antirrhinum or snapdragon (*A. majus*★), frequently escapes from cultivation and becomes established on dry sites. It has larger flowers in a variety of colours, and its calyx is 5–8 mm long, divided into broad lobes of equal length.

Digitalis purpurea★ **Plate 41**

FOXGLOVE **35–50 long**

A tall biennial or perennial with a single, erect stem topped with a conspicuous raceme of flowers, the flowers being to one side of the stalk. The leaves are broadly lance-shaped with a felted under-surface.

It lives in open, disturbed areas: flowers from October to January.

Euphrasia cuneata **Plate 41**

TUTUMAKO **15–20 long**

A medium, perennial plant with oval to broadly oval, hairless, short-stalked leaves; these have one to three pairs of blunt teeth on each side. The lower lip of the corolla is much larger than the upper lip.

It lives in open rocky locations and in scrub from sea-level to 1500 m: flowers from January to April.

Euphrasia zelandica **Plate 41**

EYEBRIGHT **6–10 long**

A low, annual plant with oval to broadly oval, hairy, stalkless leaves. The leaves have two to five narrow, pointed teeth on each side; they are clustered in rosettes at the tips of the stems, surrounding the short-stalked or stalkless flowers.

It lives in open situations, from 700 m to 1800 m: flowers from October to April.

Hebe lycopodioides Plate 42

WHIPCORD HEBE **5–8**

A tall, branching shrub, with small, closely overlapping leaves covering the branches to give the characteristic 'whipcord' appearance. Clusters of 3–12 small, white flowers, with four petals and sepals, appear at the ends of the branches. There are several whipcord species of *Hebe* with a similar appearance, but *H. lycopodioides* is the commonest.

It lives in dry, tussock grasslands and herbfields in the eastern mountains of the South Island, at 900 m to 1700 m, and flowers in December and January.

The genus *Hebe* is named after Hebe, the Greek goddess of Youth, whose chief task was to hand round nectar and ambrosia to the gods when they were feasting.

*Linaria arvensis** Plate 41

FIELD LINARIA **5 long**

A short annual with narrow, entire, stalkless leaves, the lower leaves in whorls of four. The flowers are clustered closely together at the top of the flower-stalk, and they have a white spot on the lower lip.

It grows in waste areas: flowers from July to March.

*Linaria repens** Plate 41

PALE TOADFLAX **10–16 long**

A short to medium, creeping perennial with narrow, entire stalkless leaves. The pale mauve corolla with darker veins is a distinctive feature.

It is a garden escape that is found growing in waste areas close to habitation: flowers from April to June.

Mazus radicans Plate 32

MAZUS **12–20 long**

A low, creeping perennial with usually hairy but sometimes hairless, oval, entire leaves; the dark markings on the edges of the leaf are distinctive. Up to three flowers are borne on the flower-stalks.

It grows in swamps, bogs and damp grassy areas from sea-level to 1200 m: flowers from November to March.

A related plant, *M. pumilio*, has smaller flowers (6–12 mm long) which may occasionally be blue, and occur in clusters of up to six. The name *Mazus* comes from the Greek *mazos* (teat), referring to the prominent swellings at the base of the petals.

Mimulus guttatus★ — Plate 32

MONKEY MUSK — **30–45 long**

A short, mainly hairless perennial with broad, opposite, roundly toothed leaves. The bases of the upper pairs of leaves are fused, clasping the stem. The flower is distinguished by the small red spots at its mouth, though these are sometimes absent. The throat of the flower is more or less closed by a pouch.

It is common in wet places such as swamps, stream banks and ditches: flowers from November to March.

The flower is said to resemble a monkey's face, which gives it its common name. Because it mimics the monkey, its genus name is *Mimulus*.

Mimulus moschatus★ — Plate 32

MUSK — **15–20 long**

A short, stickily hairy perennial with broad, opposite, distantly-toothed or entire leaves, and sticky hairs. The upper leaves clasp the stem. The flowers have no red spots and the throat of the flower is open.

It is common in wet places such as swamps, stream banks and ditches: flowers throughout the year.

At one time, musk was cultivated for its musky scent. In the early 20th century, unscented plants began to appear and today no scented plants are known.

Ourisia macrocarpa — Plate 42

OURISIA — **20–30**

A short, prostrate, almost hairless plant with rooting stems and erect inflorescence-stalks up to 700 mm tall. The lower surfaces of leaves, the leaf-stalks and flowering stems are often dark purple. Flowers are in whorls. The calyx lobes are broad and very bluntly tipped, sometimes slightly notched.

It grows in the South island in damp habitats from 800 m to 1300 m: flowers from October to January.

Unlike most genera of the Foxglove Family, the ourisias have almost regular flowers. The lower corolla lobe is only slightly larger than the others. A plant of similar appearance is *O. macrophylla*, but this has narrow, tapering calyx lobes, which are not notched, and has hairs on its leaf veins and often elsewhere. It lacks the purple colouration. There are 10 species of ourisia native to New Zealand and all have similar flowers. The species illustrated here are typical examples of the genus.

Ourisia sessilifolia subsp. *splendida* **Plate 42**

OURISIA **20–25**

A low, rosette-forming plant, with leaves densely covered in long, tapering (non-glandular) hairs. The leaves are rounded, with fine, rounded teeth. The hairy inflorescence-stalk bears a tight cluster of two or three flowers.

It is found in the South Island, growing in open rocky places from 1000 m to 2100 m: flowers from December to February.

The subspecies *sessilifolia* is very similar, but has spoon-shaped leaves less than 20 mm long, with long, glandular hairs also visible. The leaves are living only near the tip of the creeping stems. A closely related species, *O. simpsonii*, has the same type of leaves and hairs, except that the living leaves occur all along the stem.

Parahebe catarractae **Plate 42**

PARAHEBE **10**

A short to medium sub-shrub with oval to lance-shaped, bluntly toothed leaves 10–40 mm long. The inflorescences are long and many-flowered. The veining on the petals may be purple, as shown, or pink.

It lives in damp, rocky places, from sea-level to 500 m: flowers from November to April.

Parahebe lyallii **Plate 42**

PARAHEBE **10**

A low, creeping sub-shrub with oval to lance-shaped leaves, 5–10 mm long, with two to three blunt teeth on each side. The leaves are fleshy and usually reddish. The inflorescences are long and many-flowered.

It lives in damp, rocky places, from sea-level to 1300 m: flowers from November to March.

A similar plant, *P. decora*, has smaller leaves (1.5–4 mm long), which are entire or have a single lobe near the base. The lower three flowers of the inflorescence are in a whorl separate from the others. These plants are included as examples of the genus *Parahebe*, of which there are 11 species in New Zealand, all endemic.

Parentucellia viscosa★ **Plate 42**

TARWEED **8–12**

A short to medium annual, with an erect, unbranched stem and lance-shaped, opposite, toothed leaves, 12–45 mm long. The leaves and stem are covered with sticky hairs.

It is common in damp pastures, on roadsides and in waste areas: flowers from November to May.

This plant is semiparasitic on the roots of other plants.

Scrophularia auriculata★ **Plate 41**

WATER FIGWORT **7–10 long**

A medium to tall perennial with narrowly winged stems and leaf-stalks. Its sepals have a whitish, papery margin 1–1.5 mm wide.

It lives in damp areas on river banks and roadsides, particularly in areas of high rainfall: flowers from August to April.

Verbascum thapsus★ **Plate 43**

WOOLLY MULLEIN – AARON'S ROD **15–25**

A tall, sturdy, annual or biennial plant, covered with dense, woolly felt and with an erect, unbranched stem growing from a basal rosette. It can grow up to 3 m tall. Stem leaves are broadly oval and bluntly toothed, their bases running down to form wings on the stem. There are five stamens, three with a tuft of white hairs and two without hairs.

It lives in dry locations, beside roads, on river-beds, on poor pastures and on dry loose soils: flowers from July to April.

The name 'Aaron's rod' refers to the rod of the biblical Aaron, 'which was budded and brought forth buds and bloomed blossoms'. In early times the plant was dipped into tallow and used as a torch plant. Its large, felty leaves were used to line shoes.

Verbascum virgatum★ — Plate 43

MOTH MULLEIN — **25–40**

A tall, sturdy, biennial plant, moderately hairy, with an erect, unbranched stem growing from a basal rosette. It can grow up to 2 m tall. Stem leaves are oval, toothed, and more or less hairless. The upper stem leaves are narrower, but with lobes near the base, clasping the stem. There are five stamens, with purple or violet hairs, the two upper ones sometimes with white hairs.

It lives in open, disturbed areas: flowers from November to May.

Veronica arvensis★ — Plate 43

FIELD SPEEDWELL — **2**

A low to short, hairy, usually erect annual. The leaves are oval to rounded, irregularly toothed and short-stalked. The inflorescence is a dense, leafy, terminal raceme. In the inflorescence, the leaves are short-stalked, and are narrower and simpler than those toward the base of the plant. The petals are blue, and shorter than the sepals.

It is very common in open, modified habitats from sea-level to 1000 m: flowers from July to April.

Veronica persica★ — Plate 43

SCRAMBLING SPEEDWELL — **10–15**

A low, curly-haired, prostrate annual with short-stalked, heart-shaped leaves. The flowers are solitary, on long stalks in the axils of the leaves. The dark-veined, sky-blue flowers, with a usually white lower petal, distinguish this from other speedwells.

It is very common on open, modified areas: flowers all the year round.

The species originally lived in Persia. It was collected and grown in the botanic garden at Karlsruhe in Germany, then it escaped in about 1805 and spread throughout Europe within a few decades. Later, it spread to most other parts of the world, and now it is the commonest speedwell in New Zealand.

Veronica serpyllifolia★ — Plate 43

TURF SPEEDWELL — **8**

A low, mat-forming, almost hairless perennial with short-stalked or unstalked, almost entire, oval leaves. The flowers are solitary on stalks almost equal to the leaves. The

pale blue or white flowers, with purple veins, distinguish this from other speedwells.
It lives in damp situations, such as wet, grassy areas and beside leaves and streams: flowers all the year round.

Nightshade Family – Solanaceae

HSfewT A SiC - R
K (5) sometimes fewer or more segments
C (5)
M 5 or more; alternate
F (2)

This family provides two very important food plants: the potato and the tomato. Other food plants in the family include sweet pepper, aubergine (egg-plant), tamarillo and Cape gooseberry. The leaves of *Nicotiana tabacum* are dried and made into tobacco. Other plants, such as poroporo and deadly nightshade, are grown for the powerful alkaloids they contain, for use in medicine. Although alkaloids can be beneficial when used in the correct small dose, they are all highly poisonous substances; plants that contain them should always be regarded with caution. Alkaloid-containing plants that grow wild in New Zealand include poroporo, deadly nightshade, henbane and thorn apple. The family also has several members grown for their colourful blooms. These include cestrums, petunias, nicotianas and painted tongue (*Salpiglossis*).
Many of the above have escaped from cultivation and grow wild in New Zealand, including tomato, tobacco, petunia, Cape gooseberry, potato, cestrums, nicotianas and painted tongue.

*Datura stramonium** — Plate 40
THORN APPLE — **60–80 long**

A medium to tall annual with an unpleasant smell. The leaves are oval, with large teeth. The flowers are erect and open in the evenings, except in rainy weather. The fruits are green, oval, about 20 mm in diameter and covered with long, green thorns.
It lives in waste areas and roadsides: flowers from November to April.

It contains the hallucinogenic and very poisonous alkaloid, scopolamine. It is said that the seeds were used by the priests of Apollo to cause wild and frenzied proclamations by the Oracle.

Solanum laciniatum **Plate 40**

POROPORO **40–50**

A tall, hairless, soft-wooded shrub, the stems often with a purple tinge. The leaves are entire or deeply and pinnately cut into one to four pairs of lobes. The tips of the corolla lobes are rounded, with a distinct notch.

It grows in and around forests, plantations and hedges: flowers all the year round.

Another species known by the name of poroporo is *S. aviculare*. In this the corolla lobes are broadly oval with pointed tips, and they have no notch.

*Solanum nigrum** **Plate 40**

BLACK NIGHTSHADE **10–13**

A medium to tall annual or perennial of variable appearance. The stems are usually green, but they may be blackish. The flowers are like those of potato or tomato, with prominent, fused, yellow anthers. The fruit is a rounded berry, green at first, usually turning black as it ripens.

It grows in open areas such as roadsides and waste ground: flowers from October to May.

The plant may be poisonous, though probably only in its green parts, including the unripe fruits. Although the similarity in their common names is confusing, the highly poisonous plant, *deadly* nightshade, belongs to a different species and has a different appearance.

A plant similar to black nightshade, but with bright purple petals and yellow anthers, is bittersweet (*S. dulcamara**). Its berries are green at first, ripening to yellow, then red. It is poisonous.

Stylidium Family – Stylidiaceae

HfewS B E – I
K (5)
C (5)
M 2 sometimes 3; filaments fused into a column, surrounding the style
F (2)

A small family of herbs and a few softly wooded shrubs, mainly from Australia, Tasmania and New Zealand, it is named after the trigger plants (*Stylidium* species) of Australia. The stamens are on the column, which protrudes from the flower: this column rests to one side, but when an insect visits the flower, the column is triggered to swing across, dusting the insect's body with pollen.

Forstera bidwillii — Plate 27

FORSTERA — **6–10**

A short, creeping, hairless sub-shrub with oval, entire leaves. The main vein is narrow and inconspicuous, and the leaf has no stalk. The leaves are close together and overlapping, spreading at first, turning back when they are older. The flower-stalk bears up to three flowers. The dark red centre of the flower is a distinctive feature, though this is not always present.

It grows in grassland, herbfield and rocky places from 800 m to 1800 m: flowers from December to March.

There are three other species of *Forstera* native to New Zealand, and all are similar. *F. tenella* has thinner, often purplish leaves, less overlapping, with very short (1 mm), but definite stalks; the flower stalks are very slender, almost wire-like. It occurs in bogs from 500 m to 1400 m. The other two species have leaves with a wide, tapering, conspicuous main vein. *F. sedifolia* has broadly oval leaves 4–8 mm long, while *F. mackayi* has longer and narrower leaves 8–10 mm long.

Phyllachne colensoi — Plate 38

PHYLLACHNE — **5–7**

A low, cushion-forming perennial with thick, leathery leaves about 4 mm long and about 2 mm wide at the base, tapering to 0.5 mm wide at the tip. There is a tiny pore on the back of the leaf, near the tip. Flowers are solitary and stalkless. The column (see family description) protrudes from the flower, and bears purple anthers.

It lives in herbfield, fellfield and rocky places, from 900 m to 1900 m: flowers from November to February.

Two closely related species are *P. clavigera* (South Island only), in which the column does not or hardly protrudes from the flower, and *P. rubra*, in which the leaf tips do not taper, but are thickened into a prominent, rounded knob at the tip.

Another cushion-forming plant of similar appearance is *Donatia novae-zelandiae*; this belongs to a different family (the Donatiaceae) and is distinguished by having its two stamens *not* united into a column. (See also *Hectorella*, p. 125.)

Nasturtium Family – Tropaeolaceae

H A EC - I
K 5 usually petal-like; upper sepal is spurred
C 5 clawed
M 8
F (3)

A small family of only two genera, from South America, Mexico to Peru. Many have bright, showy blooms.

Tropaeolum majus★ — Plate 44

GARDEN NASTURTIUM — **30–40**

A trailing or scrambling annual or perennial, with distinctive, entire, peltate leaves. The spur is 25–33 mm long, more or less curved, tapering and pale orange. Petals are variously coloured and marked, most often yellow, orange or red, and double forms are sometimes found.

It lives on waste areas and roadsides, especially in damp and shady places, also on grassy areas beside coastal beaches. It flowers from October to May, but is found in flower all the year round in favourable habitats.

The leaves and flowers of a nasturtium plant growing around a wooden post reminded the botanist Linnaeus of the ancient custom of displaying the shields (=leaves) and helmets (=flowers) of a defeated enemy on a tree-trunk. The Greek name for this type of trophy was *tropaion*, from which Linnaeus derived the name *Tropaeolum*. The seeds of nasturtium are pickled and eaten with salads, in the same way as capers.

Nettle Family – Urticaceae

HST AO Si + R
P 4–5 or (4–5) sepal-like
M 4–5 opposite
F $\underline{1}$

The name 'nettle' refers to stinging nettles, not to the numerous 'dead nettles' that belong to the Lamiaceae (p. 126).

Two members of the family are grown for fibre in certain parts of the world and a few are sometimes grown for their ornamental foliage, but the family as a whole is not of economic importance.

*Urtica urens** — Plate 17

NETTLE — **0.5**

A short to medium annual, with opposite, heart-shaped leaves. Each pair of leaves is at right angles to the pair above and below, and the lower leaves are shorter than their leaf-stalks. The stem and leaves are covered with stinging hairs. The inflorescences consist of drooping racemes; flowers are either male or female, but both types occur in the same inflorescence. Male flowers have four green, slightly hairy tepals, between which can be seen four white anthers. In the female flowers, two tepals are much larger than the others. They have fine hairs on their margins, and often a single, large, stinging hair on the outer surface.

It grows in waste areas and gardens: flowers all the year round.

Valerian Family — Valerianaceae

HfewS OB Si – I
K inconspicuous, may later form a hairy pappus on the fruit
C (5) often spurred, sometimes two-lipped
M 1–4 alternate
F (3)

The stems branch repeatedly into two equal branches.
Oil of spikenard, used in perfumes, is extracted from the rhizomes of *Nardostachys jatamansi*. Cornsalad is cultivated for its leaves.

Centranthus ruber⋆ — Plate 44

SPUR VALERIAN — **4–8**

A medium to tall, greyish green perennial. The flowers are red, pink or white, the corolla is spurred at the base, and there is one anther.
It lives on banks, cliffs, roadsides and waste areas: flowers from November to June.

The leaves have a bitter taste but, if cut young, they can be eaten with salad.

Verbena Family — Verbenaceae

HSTL O E – I
K (5)
C (5) usually forming a narrow tube
M 4 alternate
F (2)

The family includes several thorny plants. The stems are usually square in section and the inflorescence is usually a spike or head of flowers, surrounded by coloured bracts.
Several timber trees are found in this family, including teak. Another member of

the family, lemon verbena, provides verbena oil, used in making perfumes. Ornamental plants in this family include several species of *Verbena* and *Lantana*.

Lantana camara★ — Plate 44

LANTANA — **10 long**

A tall, erect or scrambling shrub with a distinctive smell. Its stems have backwardly directed prickles. The two-coloured inflorescence is distinctive, as flowers change colour from cream or pale yellow to pink or orange as they mature.

It lives on roadsides and waste areas, and flowers throughout the year.

Lantana lives in many other countries, mainly in the tropics and sub-tropics, and is regarded as being one of the most troublesome weeds in the world. A related species is trailing lantana (*L. montevidensis*★), which has stems only up to 200 mm tall, and pink to mauve flowers.

Verbena bonariensis★ — Plate 44

PURPLE-TOP — **3**

A tall, erect perennial, its stems square with bristly hairs on the angles. The leaf base has two lobes, partly clasping the stem. Flowers are small with pink-purple petals, arranged in dense spikes up to 30 mm long, at the top of the plant. The spikes do not elongate appreciably at fruiting.

It lives on roadsides and other open waste areas, and also in pastures: flowers from January to June.

Its species name means 'from Buenos Aires', the city in South America. A related, though less common, species is *V. brasiliensis*★, from Brazil; the margins of its leaves taper gradually to the point of attachment to the stem, instead of being lobed.

Verbena litoralis★ — Plate 44

BLUE VERVAIN — **2–3**

A tall, square-stemmed perennial, with a rough feeling on the angles of the stem. The leaves are mostly stalked, except the upper ones. They are lance-shaped to rhomboid, toothed, and with sharp hairs. The spikes become longer (more than 50 mm long) as the flowers develop, and the flowers are not as densely packed as in purple-top. They elongate to 150 mm long at fruiting.

It grows on roadsides and track-sides and other open waste areas, especially near the coast: flowers all through the year.

Vervain (*V. officinalis**) is similar, but has pinnate or pinnately lobed leaves and spikes up to 150 mm long at maximum flowering, and 250 mm long at fruiting.

Violet Family – Violaceae

HST BA Si + IR
K 5
C 5 lower petal may be spurred, containing a nectary
M 5 alternate
F (3)

Although the family is well known for its violets and pansies, its only plant of economic importance is *Viola odorata*, from which oils used in perfumes and flavourings are obtained.

Viola cunninghamii — Plate 32

VIOLA — **10–20**

A low to short perennial with no stem, the leaves and flower-stalks arising from a many-headed rootstock. Leaves are broadly oval on long (20–100 mm), flat leaf-stalks; the leaf margin is cut straight across at the base, or tapers gradually into the leaf-stalk. The margins have 5–10 very shallow, rounded teeth on each side.

It lives in damp sites from sea-level to 1800 m: flowers from December to March.

Two other native violets with white, purple-veined petals have leafy stems. *V. filicaulis* has a creeping, rooting stem and rounded to kidney-shaped leaves on leaf-stalks 10–20 mm long. *V. lyallii* has a prostrate stem with heart-shaped to kidney-shaped leaves on stalks 20–50 mm long.

Appendix

IDENTIFYING FAMILIES BY COMPUTER

This computer program is used for identifying dicot families. It uses the same table of features as on p. 45, but, since the computer does the matching, it is quicker and less liable to error. A listing of the program for the Amstrad CPC series computers is given below. As far as possible, the BASIC keywords we have used are found in all versions of BASIC, so you should have no problems in adapting this program to other computers, such as the Commodore 64 and the Spectrum. The program is very short, and most of the typing consists of keying in the lines of data that specify the families.

```
10 REM *** IDENTIFY DICOT FAMILIES ***
20 MODE 2
30 DIM family$(53), feature$(53),confirm$(53)
40 FOR j=1 TO 53:READ family$(j):READ feature$(j
):READ confirm$(j):NEXT
50 plant$="":PRINT"Answer y, n, or x"
60 PRINT"Leaves opposite";
70 GOSUB 410
80 PRINT"Leaves compound";
90 GOSUB 410
100 PRINT"Stipules present";
```

```
110 GOSUB 410
120 PRINT"Flowers irregular";
130 GOSUB 410
140 PRINT"Sepals united";
150 GOSUB 410
160 PRINT"Petals united";
170 GOSUB 410
180 PRINT"Stamens many";
190 GOSUB 410
200 PRINT"Ovary inferior";
210 GOSUB 410
220 PRINT "Carpels free";
230 GOSUB 410
240 PRINT
250 PRINT"The family is:"
260 family=1:found=0
270 feature=1
280 IF MID$(feature$(family),feature,1)="b" THEN
 320
290 IF MID$(plant$,feature,1)="x" THEN 320
300 IF MID$(feature$(family),feature,1)=MID$(pla
nt$,feature,1) THEN 320
310 GOTO 350
320 IF feature<9 THEN feature=feature+1:GOTO 280
330 PRINT family$(family)+"aceae - ";confirm$(fa
mily)
340 found=found+1
350 IF family<53 THEN family=family+1:GOTO 270

360 IF found=0 THEN PRINT:PRINT"No family found"
370 PRINT:PRINT"Key 'c' to continue"
380 IF INKEY$<>"c" THEN 380
390 CLS:GOTO 50
400 REM ******************************
410 INPUT answer$
420 IF answer$="y" OR answer$="n" OR answer$="x"
 THEN plant$=plant$+answer$:RETURN
430 GOTO 410
440 REM ******************************
450 DATA Aizo,ynnnnnyyn,fleshy leaves; many peta
ls and sepals
460 DATA Amaranth,bnnnnnnnn,small chaffy fls;T3-
5;M5 opposite C
470 DATA Api,nybnnnnyn,umbels; aromatic;K5;C5;M5
```

```
480 DATA Apocyn,ynnnnynnn,twining; long pods;M5
490 DATA Aster,bbnbnynyn,fl-heads; M5 fused;K is
 pappus
500 DATA Boragin,nnnnyynnn,bristly-hairy; fl ste
m curled; 4 nutlets
510 DATA Brassic,nbnnnnnnn,cross-flowers;M 4-lon
g + 2-short OR 4
520 DATA Campanul,nnnnyynyn,usually blue or viol
et;fls bell-shaped
530 DATA Caprifoli,ynnbyynyn,shrubs; woody climb
ers
540 DATA Caryophyll,ynbnbnnnn,stem nodes swollen
; M twice C
550 DATA Chenopodi,nnnnynnnn,inconspicuous fls;
not chaffy;pl mealy
560 DATA Clusi,ynnnnnynn,stamen bundles;dotted l
vs & K;yellow fls
570 DATA Convolvul,nnnnnynnn,twining stems; funn
el shaped fls; latex
580 DATA Crassul,ynnnbbnny,fleshy; F same no as
C
590 DATA Dipsac,ybnyyynyn,fl-heads;M4
600 DATA Droser,nnnnynnnn,long glandular hairs;
insectivores
610 DATA Eric,bnnnyynnn,low shrubs; M twice C
620 DATA Euphorbi,bnynnnnnn,latex; fls unisexual
;C=0
630 DATA Fab,nyyyynnnn,'pea' fls;pods
640 DATA Fumari,nynynnnnn, stems brittle;petal s
purs
650 DATA Gentian,ynnnyynnn,sessile lvs; bell-sha
ped fls; hairless
660 DATA Gerani,bnybbnnnnn,'stork's-bill' fruits

670 DATA Goodeni,nnnyyynyn,pollen cup around sti
gma;C tube split
680 DATA Halorag,bnnnynnyn,often aquatic;fruits
hard and squarish
690 DATA Hectorell,nnnnnnnnn,cushion-pl;K2;C5;M5
;(Hectorella)
700 DATA Lami,ynnyyynnn,square stems; aromatic;
4 nutlets
710 DATA Lin,bnnnnnnnn,unbranched erect stems; s
mall narrow lvs
```

```
720 DATA Lobeli,nnnyyynyn,M joined in a tube;C t
ube split
730 DATA Lythr,ynnnynnnn,epicalyx; calyx-tube
740 DATA Malv,nnynynynn,lobed leaves;M filaments
 joined
750 DATA Onagr,bnnnynnyn,calyx-tube;M 4 or 8; K4
; C4
760 DATA Orobanch,nnnyyynnn,no chlorophyll; para
sitic; scale lvs
770 DATA Oxalid,nyynnnnnn,trefoil leaves;orange
calli lvs & spls
780 DATA Papaver,nbnnbnynn,K 2; falling off in b
ud;latex
790 DATA Passiflor,nnynnnnnn,corona;tendrils

800 DATA Phytolacc,nnnnynnny,small fls in dense
spikes;C=0
810 DATA Plantagin,nnnnyynnn,basal lvs; parallel
 vns;small chaffy fls
820 DATA Polemoni,bbnnyynnn,K5;C5;M5
830 DATA Polygon,nnynnnnnn,ochreae; small fls;sw
ollen stem nodes
840 DATA Portulac,bnnnbnnnn,fleshy leaves; satin
y petals;M2;K irreg
850 DATA Primulac,ynnnyynnn, leaves basal; M opp
osite C
860 DATA Ranuncul,bbnnnnyny,P sometimes of 1 who
rl; M in spiral
870 DATA Resed,nnnynnynn,K 4-6; C 4-6; M 11-25

880 DATA Ros,nyynnnyny,epicalyx;M & F many (exce
pt Acaena)
890 DATA Rubi,ynynnynyn,leaves appear whorled
900 DATA Scrophulari,bnnnyynnn,M fewer than C;F
2
910 DATA Solan,nbnnyynnn,'spire' of fused anther
s;nasty smell
920 DATA Stylidi,nnnnyynyn,M 2 united into colum
n
930 DATA Tropaeol,nbnynnnnn,lvs peltate;sepal sp
ur;clawed petals
940 DATA Urtic,bnynnnnnn,stinging hairs; unisexu
al fls
950 DATA Valerian,ynnynynyn,dichotomous branchin
```

```
g;M 1-4;petal spur
960 DATA Verben,ynnyyynnn,square stems;thorny;na
rrow C tube
970 DATA Viol,nnyynnnnn,petal spurs;M fused in r
ing;solitary fls
```

Keying in

If you have an Amstrad CPC series computer, key in the listing exactly as given here, and save it. This listing is printed out directly from the computer and will work straight away. If you have a different computer, most lines will be the same as those shown in the listing. The following points may well help you make the few conversions that are necessary:

- Line 20 puts the computer into 80-column mode. If your computer does not have this mode (for example, Spectrum) do not worry; the displayed lines will simply be carried on to the line below. In the Amstrad, this line automatically clears the screen.
- In line 30 there are three arrays: *family$* holds the names of the families, without the ending '-aceae'; *feature$* holds strings consisting of y, n and b, as in the table on p. 45; *confirm$* holds the confirmatory details, as in the table. Each array has an entry for each of the 53 dicot families.
- Lines 280–300: you may need to alter the MID$ expressions, but this is unlikely because this format is standard for most versions of BASIC.
- Line 380: this waits for key 'c' to be pressed before continuing. The INKEY$ statement may need adapting, as this works slightly differently in a few versions of BASIC.
- Line 390: CLS clears the screen.
- Lines 450–970: the data consists of the family name, the features and the confirmatory features. You may need to put quotation marks (") around the data items in some forms of BASIC. Beware of using commas in the features lists; in BASIC, commas separate one item of data from another, which is why we have used semicolons in the features lists.

Using the program

Make sure that the computer is in lower-case mode before you run the program. As the questions appear on the screen, answer them by keying 'y', 'n' or 'x' and pressing RETURN (or ENTER). Key 'x' when you are not sure what answer to give. When you have answered all questions, the computer searches the table and prints out the names of the families that match the answers you have typed in. It also prints the confirmatory features. Sometimes more than one family matches, in which case use the confirmatory features to decide which of these families is the one. When you have found the family, turn to the appropriate section of the book to discover the genus and species. If you have more plants to identify, key 'c' and the program is repeated.

Remember that the table is specially designed for identifying those species which are described in this book. Though it very often works for other species as well, it is not guaranteed to do so; if the answers you give do not match any of the families described in the table, the message 'No family found' appears.

PHOTOGRAPHIC NOTES

The photographs in this book were taken to enable plants to be easily identified; it was essential therefore to depict colour, form and detail clearly and correctly. Kodachrome 64 and 25 films were used throughout. Photographs were taken with an Olympus OM2 camera fitted with a 50 mm or 80 mm Zuiko macro lens.

Light is perhaps the most important element of flower photography, and this should be fairly strong but diffuse. Ideal results are obtained on thinly overcast days when the sun throws the faintest of shadows. Full sunlight can render the subtle form and colour of a flower in a hideous mixture of burnt-out highlights and hopelessly dark shadows. At the other end of the scale, heavily overcast days can provide flat, dimensionless light and often require the use of shutter speeds that are so long that the film fails to register colour correctly.

In order to photograph a flower with all its parts in focus, a reasonably small lens aperture is required. Under diffuse, overcast conditions this invariably means using a shutter speed which is too slow for hand holding. To avoid any camera shake, a solid tripod is therefore indispensable. The only enemy which then remains is the wind: even the slightest breeze will send the flower into a spasm of nodding and the photographer into a fit of despair. A calm disposition and some ingenuity both help to improvise a method to steady or shelter the flower.

In situations where natural light was not suitable to show the flower to best effect, flashlight was used; two flash heads were used to provide even lighting, and these were adjusted to accentuate surface or other detail as necessary.

ACKNOWLEDGEMENTS

There are seven authors without whose works we should never have been able to undertake or complete this book. These are the authors of the four official floras of New Zealand:

Allan, H. H., *Flora of New Zealand*, Volume 1, 1961.
Moore, L. B., and Edgar, E., *Flora of New Zealand*, Volume II, 1976
Healy, A. J. and Edgar, E., *Flora of New Zealand*, Volume III, 1980
Webb, C. J., Sykes, W. R., and Garnock-Jones, P. J., *Flora of New Zealand*, Volume IV, 1988

The first three volumes listed above are published by the Government Printer, Wellington, and the fourth by the Botany Division of DSIR, Christchurch. As well as these, we have made reference to the works of the following authors:

Addison, Josephine, *The Illustrated Plant Lore*, Guild Publishing, London, 1985

Dahlgren, R. M. T., Clifford, H. T., and Yeo, P. F., *The Families of the Monocotyledons*, Springer-Verlag, Berlin, 1985

Fitter, Richard, Fitter, Alastair, and Blamey, Marjorie, *The Wild Flowers of Britain and Northern Europe*, 4th edition, Collins, London, 1985

Genders, Roy, *The Scented Wild Flowers of Britain*, Collins, London, 1971

Gordon, Lesley, *Poorman's Nosegay*, Collins and Harvill Press, London, 1973

Mark, A. F., and Adams, Nancy M., *New Zealand Alpine Plants*, 2nd edition, Reed Methuen, Auckland, 1986

Martin, W. Keble, *The Concise British Flora in Colour*, Rainbird, London, 1965

Moore, L. B., and Irwin, J. B., *The Oxford Book of New Zealand Plants*, Oxford University Press, Auckland, 1978

Salmon, John, *Alpine Plants of New Zealand*, 2nd edition, Collins, Auckland, 1985

We thank Peter Johnson for allowing us to use his photograph of scabweed. We would also like to thank Newmans Rentals Limited, Auckland, for assisting us with a camper-van in which to undertake some of our searches for plants.

FURTHER READING

The best books for identifying the whole range of New Zealand flowering plants (except the grasses) are the first four floras listed under 'Acknowledgements'. For monocots, it is best to begin by using the keys in Volume III, which cover both native and introduced genera. If the plant belongs to a native genus, its species can be determined by using the keys in Volume II. Similarly, for dicots it is best to begin with the keys in Volume IV, referring later to Volume I to identify the species of native genera.

Popular books on alpine plants include those by Salmon, and by Mark and Adams, also listed under 'Acknowledgements'; these are fully illustrated and are easy to use.

For the identification of grasses, a helpful book is *What Grass Is That?*, by N. C. Lambrechtsen; this deals with introduced grasses and is illustrated by line drawings. A booklet that includes many of the commonest trees is the *Field Guide to Common New Zealand Trees and Shrubs*, published in 1982 by the New Zealand Forest Service and the Department of Lands and Survey. It is illustrated by line drawings and includes a key. Both of these books are available from the Ministry of Agriculture and Fisheries at their Primedia Bookshop, Wellington. Rather more comprehensive is *A Field Guide to the Native Trees of New Zealand*, by J. T. Salmon, published by Reed Methuen, Auckland, 1986; this is fully illustrated by colour photographs of high quality.

Index

Plants are indexed under common names and genus names (common names are indexed *without* the epithets 'common' and 'wild'). Families are indexed under common names and scientific names.